T0140530

Studies in Big Data

Volume 20

Series editor

Janusz Kacprzyk, Polish Academy of Sciences, Warsaw, Poland
e-mail: kacprzyk@ibspan.waw.pl

About this Series

The series "Studies in Big Data" (SBD) publishes new developments and advances in the various areas of Big Data- quickly and with a high quality. The intent is to cover the theory, research, development, and applications of Big Data, as embedded in the fields of engineering, computer science, physics, economics and life sciences. The books of the series refer to the analysis and understanding of large, complex, and/or distributed data sets generated from recent digital sources coming from sensors or other physical instruments as well as simulations, crowd sourcing, social networks or other internet transactions, such as emails or video click streams and other. The series contains monographs, lecture notes and edited volumes in Big Data spanning the areas of computational intelligence incl. neural networks, evolutionary computation, soft computing, fuzzy systems, as well as artificial intelligence, data mining, modern statistics and Operations research, as well as self-organizing systems. Of particular value to both the contributors and the readership are the short publication timeframe and the world-wide distribution, which enable both wide and rapid dissemination of research output.

More information about this series at http://www.springer.com/series/11970

Oliver Kramer

Machine Learning
for Evolution Strategies

 Springer

Oliver Kramer
Informatik
Universität Oldenburg
Oldenburg
Germany

ISSN 2197-6503   ISSN 2197-6511   (electronic)
Studies in Big Data
ISBN 978-3-319-81500-8   ISBN 978-3-319-33383-0   (eBook)
DOI 10.1007/978-3-319-33383-0

Printed on acid-free paper

This Springer imprint is published by Springer Nature
The registered company is Springer International Publishing AG Switzerland

Contents

Abstract

Evolution strategies are successful techniques for continuous blackbox optimization. Many applications have proven that they are excellent means for solving practical problems. Machine learning comprises a rich set of algorithms for learning from data with an emphasis on pattern recognition and statistical learning. This book introduces numerous algorithmic hybridizations between both worlds that show how machine learning can improve and support evolution strategies. The set of methods comprises covariance matrix estimation, meta-modeling of fitness and constraint functions, dimensionality reduction for search and visualization of high-dimensional optimization processes, and clustering-based niching. After giving an introduction to evolution strategies and machine learning, the book builds the bridge between both worlds with an algorithmic and experimental perspective. Experiments mostly employ a (1+1)-ES and are implemented in PYTHON using the machine learning library SCIKIT-LEARN. The examples are conducted on typical benchmark problems illustrating algorithmic concepts and their experimental behavior. The book closes with a discussion in related lines of research.

Chapter 1
Introduction

1.1 Computational Intelligence

Computational intelligence is an increasingly important discipline with an impact on more and more domains. It mainly comprises the two large problem classes optimization and machine learning that are strongly connected to each other, but which also cover a broad field of individual problems and tasks. Examples for applications of optimization and learning methods range from robotics to civil engineering. Optimization is the problem of finding optimal parameters for all kinds of systems. Machine learning is an essential part of intelligence. It means that a system is able to recognize structure in data and to predict the future based on past observations from the past. It is therefore related to human learning and thinking. The benefit of learning is the ability to behave reasonable based on experience, in particular considering functional aspects that can be extracted from observations. Machine learning is basically an advancement of statistical methods, a reason why it also denoted as statistical learning.

Traditionally, computational intelligence consists of three main branches: (1) evolutionary algorithms, (2) neural networks, and (3) fuzzy logic. Evolutionary algorithms comprise methods that allow the optimization of parameters of a system oriented to the biological principle of natural evolution. Neural networks are also biologically inspired methods. They are abstract imitations of natural neural information processing. Fuzzy logic allows modeling of linguistic terms with linguistic inaccuracy. Meanwhile, numerous other methods have been developed and influence the three branches of computational intelligence.

1.2 Optimization

Optimization is an important problem class in computer science finding numerous applications in domains like electrical engineering, information management, and many more. Optimization variables may be numerical, discrete, or combinatorial.

© Springer International Publishing Switzerland 2016
O. Kramer, *Machine Learning for Evolution Strategies*,
Studies in Big Data 20, DOI 10.1007/978-3-319-33383-0_1

Fig. 1.1 Illustration of local
and global optima of a
minimization problem

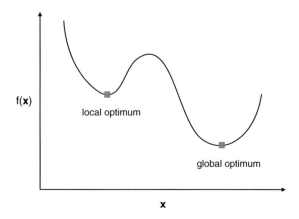

Moreover, mixed-integer problems contain continuous and discrete variables. A famous combinatorial optimization problem is the traveling salesman problem (TSP), which is the task to find the shortest tour between a set of cities, which have to be visited. This problem can be arbitrarily difficult. For example, if we consider 100 cities, the number of possible tours exceeds 10^{150}, which is nearly impossible to compute. The search in this large solution space is a particularly difficult problem. But the world of combinatorial optimization profits from developments in computational complexity. This theory allows differentiating between easy and hard problems. For example, the TSP problem is NP-hard, which means that (probably—to the best of our current knowledge) no algorithm exists that computes a solution in polynomial time.

In this work, we focus on continuous optimization also referred to as numerical optimization. We define an optimization problem as the minimization of an objective function $f(\mathbf{x})$ w.r.t. a set of d parameters $\mathbf{x} = (x_1, \ldots, x_d)^T \in \mathbb{R}^d$. Every minimization problem can easily be treated as maximization problem by setting $f' = -f$. Figure 1.1 illustrates the definition of optimality. A local optimum employs the best fitness in its neighborhood, a global optimum employs the best overall fitness. A global optimum is also a (special) local optimum. Optimization methods must prevent getting stuck in local optima.

Some optimization problems are high-dimensional, i.e., a large number of parameters has to be optimized at the same time, while others only concern few variables. High-dimensional optimization problems often occur during computer experiments, when a complex simulation model replaces a lab experiment. The problem that comes up under such conditions is that due to the curse of dimensionality problem no reasonable search in the solution space is possible. In such cases, a stochastic trial-and-error method can yield good results. This also holds for problems that are difficult due to multiple local optima or non-smooth and noisy fitness landscapes.

Many conditions can make an optimization problem difficult to solve:

- Often, no analytic expression is available. In this case, it is not possible to apply methods like gradient descent, Newton methods, and other variants.
- Problems may suffer from noise, which means that the fitness function evolution of the same candidate solution may result in different fitness values.
- Solution spaces may be constrained, i.e., not the whole solution space is feasible. As optima often lie at the constraint boundary, where the success rate drops significantly, heuristics like evolutionary algorithms must be enhanced to cope with changing conditions.
- When more than two objectives have to be optimized at the same time, which may be conflictive, a Pareto set of solutions has to be generated. For this sake, evolutionary multi-objective optimization algorithms are a good choice as they naturally work with populations, which allow the evolution of a Pareto set.

Under such conditions classic optimization strategies often fail and heuristics perform good results. Evolutionary algorithms (EAs) are excellent methods to find optima for blackbox optimization problems. EAs are inspired by the natural process of evolution. Different branches began to develop in the mid of the last century. In Germany, Rechenberg and Schwefel [1–3] developed evolution strategies (ES) in the sixties in Berlin. A similar development took place in the United States initiated by Holland [4] at the same time. Methods developed in this line of research were called genetic algorithms. All EA research branches have grown together in the recent decade, as similar heuristic extensions can be used and exchanged within different EA types.

1.3 Machine Learning and Big Data

Machine learning allows learning from data. Information is the basis of learning. Environmental sensors, text and image data, time series data in economy, there are numerous examples of information that can be used for learning. There are two different types of learning: supervised and unsupervised. Based on observations, the focus of supervised learning is the recognition of functional relationships between patterns. Labels are used to train a model. Figure 1.2 illustrates a supervised learning scenario. In classification, a new pattern $\mathbf{x}'$ without label information is assigned to a label based on the model that is learned from a training data set. Many effective and efficient machine learning methods have been proposed in the past. Strong machine learning libraries allow their fast application to practical learning problems. Famous methods are nearest neighbor methods, decision trees, random forests, and support vector machines. Deep learning is a class of methods that recently attracts a lot of attention, in particular in image and speech recognition.

In unsupervised learning scenarios, the idea is to learn from the data distribution without further label information. Clustering, dimensionality reduction, and outlier detection are the most important unsupervised learning problems. Clustering is the task to learn groups (clusters) of patterns, mostly based on the data distributions

Fig. 1.2 Illustration of
supervised learning with the
nearest neighbor classifier,
see Chap. 6. We seek for the
label of a novel pattern $\mathbf{x}'$

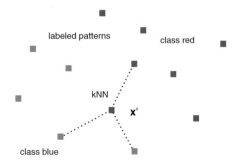

and on data densities. Dimensionality reduction is a further typically unsupervised[1] problem class that maps patterns from high-dimensional spaces to spaces with lower dimensionality. Applications are visualization of high-dimensional data spaces and pre-processing of supervised learning methods. Famous examples for unsupervised learning methods are k-means, Spatial-Density Clustering of Applications with Noise (DBSCAN), isometric mapping, and self-organizing maps.

Besides the development and analysis of new methods, their theoretical investigation is an important aspect. Theory helps to understand aspects like runtime behavior and convergence properties. It often requires the reduction of problems and methods to simplified variants that are not applied in practice. However, theoretical investigations often reveal the general aspects of problem hardness and method complexity that are required to understand.

The application aspect plays a very important role in machine learning. To improve the applicability of machine learning concepts strong machine learning libraries have been developed. An example is the SCIKIT-LEARN library that will exhaustively be used in this book. A deeper knowledge about the problem domain usually helps to adapt and improve the machine learning process chain. The optimization of this process chain comprises the selection of appropriate methods for aggregating data sets, pre-processing, learning, post-processing, and visualization of results. Deeper knowledge about various technologies is required in practice like heterogeneous data bases, low-level programming languages to access embedded sensors, network technologies for the collection of data and distribution of computations, and computing architectures like multi-processor systems and distributed computing settings.

With Big Data, we usually associate the situation of very large data sets, from which we want to learn. This scenario often imposes difficult challenges on various aspects of the machine learning process chain. An optimization of the tool chain is often required. If billions of patterns have to be processed in a short time this affords the scaling of the computing architecture or the distribution of processes and tasks on many machines. For the machine learning expert, the proper development and choice of efficient machine learning methods with low runtime complexity but large effectivity is the most important task.

---

[1] Also supervised dimensionality reduction methods exist.

1.4 Motivation

The objective of this book is to show that the intersection of both worlds, evolutionary computation and machine learning, lead to efficient and powerful hybridizations. Besides the line of research of evolutionary computation for machine learning, the other way around, i.e., machine learning for evolutionary computation, is a fruitful research direction. The idea is to support the evolutionary search with machine learning models, while concentrating on the specific case of continuous solution spaces. Various kinds of hybridizations are possible reaching from covariance matrix estimation to meta-modeling of fitness and constraint functions. Numerous challenges arise when both worlds are combined with each other.

This book is an introductory depiction building a bridge between both worlds from an algorithmic and experimental perspective. Experiments on small sets of benchmark problems illustrate algorithmic concepts and give first impressions of their behaviors. There is much space for further investigations, e.g., for theoretical bridges that help to understand the interplay between methods. For the future, we expect a lot of further kinds of ways to improve evolutionary search with machine learning techniques.

The book only requires basic knowledge in linear algebra and statistics. Nevertheless, the reader will find easy descriptions and will not have to dig through endless formalisms and equations. The book can be used as:

- introduction to evolutionary computation,
- introduction to machine learning, and
- guide to machine learning for evolutionary computation.

The book gives an introduction to problem solving and modeling, to the corresponding machine learning methods, and overviews of the most important related work, and short depictions of experimental analyses.

1.5 Benchmark Problems

The analysis of algorithms in the intersection of machine learning and evolutionary computation are most often based on computer experiments. Computer experiments are the main analytic tool of this work. In blackbox optimization, such experimental analyses are focused on particular benchmark functions with characteristic properties, on which the analyzed algorithms should prove to succeed. Various benchmark function have been proposed in literature. In this work, we only use a small set of benchmark problems, as the main focus is to demonstrate potential hybridizations. Of course, experimental analyses are always limited to some extend. On the one hand, the concentration on particular benchmark problems is the only possible way of experimentalism as it will never be possible to analyze the algorithmic behavior on all kinds of problems. On the other hand, the concentration can lead to a bias towards a restricted set of problems.

The focus of this book to a small set of known benchmark problem is motivated by the fact that the employed machine learning algorithms have already proven to be valid and efficient methods in various data mining and machine learning tasks. The task of this book is to demonstrate, how they can be integrated into evolutionary search and to illustrate their behaviors on few benchmark functions.

Computer experiments are the main methodological tool to evaluate the algorithms and hybridizations introduced in this work. As most algorithms employ randomized and stochastic components, each experiment has to be repeated multiple times. The resulting mean value, median and the corresponding standard deviation give an impression of the algorithmic performance. Statistical tests help to evaluate, if the superiority of any algorithm is significant. As the results of most EA-based experiments are not Gaussian distributed, the standard student T-test can usually not be employed. Instead, the rank-based tests like the Wilcoxon test allow reasonable conclusions. The Wilcoxon signed-rank test is the analogue to the T-test. The test makes use of the null hypothesis, which assumes that the median difference between pairs of observations is zero. It ranks the absolute value of the differences between observations from the smallest (rank 1) to the largest. The idea is to add the ranks of all differences in both directions, while the smaller sum is the output of the test.

1.6   Overview

This book is structured in ten chapters. The following list gives an overview of the chapters and their contributions:

Chapter 2

In Chap. 2, we introduce the basic concepts of optimization with evolutionary algorithms. It gives an introduction to evolutionary computation and nature-inspired heuristics for optimization problems. The chapter illustrates the main concepts of translating evolutionary principles into an algorithmic framework for optimization. The (1+1)-ES with Rechenberg's mutation strength control serves as basis of most later chapters.

Chapter 3

Covariance matrix estimation can improve ES based on Gaussian mutations. In Chap. 3, we integrate the Ledoit-Wolf covariance matrix estimation method into the Gaussian mutation of the (1+1)-ES and compare to variants based on empirical maximum likelihood estimation.

Chapter 4

In Chap. 4, we sketch the main tasks and problem classes in machine learning. We give an introduction to supervised learning and pay special attention to the topics model selection, overfitting, and the curse of dimensionality. This chapter is an important introduction for readers, who are not familiar with machine learning modeling and machine learning algorithms.

Chapter 5

Chapter 5 gives an introduction to SCIKIT-LEARN, a machine learning library for PYTHON. With a strong variety of efficient flexible methods, SCIKIT-LEARN implements algorithms for pre-processing, model selection, model evaluation, supervised learning, unsupervised learning, and many more.

Chapter 6

In Chap. 6, we show that a reduction of the number of fitness function evaluations of a $(1+1)$-ES is possible with a combination of a k-nearest neighbor (kNN) regression model, a local training set of fitness function evaluations, and a convenient meta-model management. We analyze the reduction of fitness function evaluations on a small set of benchmark functions.

Chapter 7

In the line of research on fitness function surrogates, the idea of meta-modeling the constraint boundary is the next step we analyze in Chap. 7. We employ support vector machines (SVMs) to learn the constraint boundary as binary classification problem. A new candidate solution is first evaluated on the SVM-based constraint surrogate. The constraint function is only evaluated, if the solution is predicted to be feasible.

Chapter 8

In Chap. 8, we employ the concept of bloat by optimizing in a higher-dimensional solution space that is mapped to the real-solution space with dimensionality reduction, more specifically, with principal component analysis (PCA). The search in a space that employs a larger dimensionality than the original solution space may be easier. The solutions are evaluated and the best w.r.t. the fitness in the original space are inherited to the next generation.

Chapter 9

The visualization of evolutionary blackbox optimization runs is important to understand evolutionary processes that may require the interaction with or intervention by the practitioner. In Chap. 9, we map high-dimensional evolutionary runs to a two-dimensional space easy to plot with isometric mapping (ISOMAP). The fitness of embedded points is interpolated and visualized with `matplotlib` methods.

Chapter 10

In multimodal optimization, the task is to detect most global and local optima in solution space. Evolutionary niching is a technique to search in multiple parts of the solution space simultaneously. In Chap. 10, we present a niching approach based on an explorative phase of uniform sampling, selecting the best solutions, and applying clustering to detect niches.

Chapter 11

In Chap. 11, we summarize the most important findings of this work. We give a short overview to evolutionary search in machine learning and give insights into prospective future work.

1.7 Previous Work

Parts of this book built upon previous work that has been published in peer-reviewed conferences. An overview of previous work is the following:

- The covariance matrix estimation approach of Chap. 3 based on Ledoit-Wolf estimation has been introduced on the Congress on Evolutionary Computation (CEC) 2015 [5] in Sendai, Japan.
- The nearest neighbor meta-model approach for fitness function evaluations presented in Chap. 6 is based on a paper presented at the EvoApplications conference (as part of EvoStar 2016) [6] in Porto, Portugal.
- The PCA-based dimensionality reduction approach that optimizes in high-dimensional solution spaces presented in Chap. 8 has been introduced on the International Joint Conference on Neural Networks (IJCNN) 2015 [7] in Killarney, Ireland.
- The ISOMAP-based visualization approach of Chap. 9 has also been introduced on the Congress on Evolutionary Computation (CEC) 2015 [8] in Japan.

Parts of this work are consistent depictions of published research results presenting various extended results and descriptions. The book is written in a scientific style with the use of "we" rather than "I".

1.8 Notations

We use the following notations. Vectors use small bold latin letters like $\mathbf{x}$, scalars small plain Latin or Greek letters like σ. In optimization scenarios, we use the variable $\mathbf{x} = (x_1, \ldots, x_d)^T \in \mathbb{R}^d$ for objective variables that have to be optimized w.r.t. a fitness function f. In machine learning, concepts often use the same notation. Patterns are vectors $\mathbf{x} = (x_1, \ldots, x_d)^T$ of attributes or features x_j with $j = 1, \ldots, d$ for machine learning models f. This is reasonable as candidate solutions in optimization are often treated as patterns in this book. Patterns lie in an d-dimensional data space and are usually indexed from 1 to N, i.e., a data set consists of patterns $\mathbf{x}_1, \ldots, \mathbf{x}_N$. They may carry labels $y_1, \ldots, y_N$ resulting in pattern-label pairs

$$\{(\mathbf{x}_1, y_1), \ldots, (\mathbf{x}_N, y_N)\}.$$

When the dimensionality of objective variables or patterns differs from d (i.e., from the dimensionality, where the search actually takes place), we employ the notation $\hat{\mathbf{x}}$. In Chap. 8, $\hat{\mathbf{x}}$ represents an abstract solution of higher dimensionality, in Chap. 9 it represents a visualizable low-dimensional pendant of solution $\mathbf{x}$.

While in literature, f stands for a fitness function in optimization and for a supervised model in machine learning, we resolve this overlap in the following. While being used this standard way in the introductory chapters, Chap. 6 uses f for the fitness function and $\hat{f}$ for the machine learning model, which is a surrogate of f. In Chap. 7, f is the fitness function, while g is the constraint function, for which a machine learning surrogate $\hat{g}$ is learned. In Chaps. 8 and 9, f is the fitness function and F is the dimensionality reduction mapping from a space of higher dimensions to a space of lower dimensionality. In Chap. 10, f is again the fitness function, while c delivers the cluster assignments of a pattern.

A matrix is written in bold large letters, e.g., $\mathbf{C}$ for a covariance matrix. Matrix $\mathbf{C}^T$ is the transpose of matrix $\mathbf{C}$. The p-norm will be written as $\| \cdot \|_p p$ with the frequently employed variant of the Euclidean norm written as $\| \cdot \|_2$.

1.9 Python

The algorithms introduced in this book are based on PYTHON and built upon various well-known packages including `Numpy` [9], `SciPy` [10], `Matplotlib` [11], and SCIKIT-LEARN [12]. PYTHON is an attractive programming language that allows functional programming, classic structured programming, and objective-oriented programming. The installation of PYTHON is usually easy. Most Linux, Windows, and Mac OS distributions already contain a native PYTHON version. On Linux systems, PYTHON is usually installed in the folder `/usr/local/bin/python`. If not, there are attractive PYTHON distributions that already contain most packages that are required for fast prototyping machine learning and optimization algorithms. First experiments with PYTHON can easily be conducted when starting the PYTHON interpreter, which is usually already possible, when typing PYTHON in a Unix or Windows shell.

An example for a typical functional list operation that demonstrates the capabilities of PYTHON is:

```
[f(x)*math.pi for x in range(1,10)]
```

This expression generates a list of function values with arguments $1, \ldots, 10$ of function `f(x)`, which must be defined before, multiplied with π.

A very frequently employed package is `Numpy` for scientific computing with PYTHON. It comprises powerful data structures and methods for handling N-dimensional array objects and employs outstanding linear algebra and random number functions. `SciPy` provides efficient numerical routines, e.g., for numerical integration, for optimization, and distance computations. `Matplotlib` is a

PYTHON plotting library that is able to generate high-quality figures. SCIKIT-LEARN is a PYTHON framework for machine learning providing efficient tools for data mining and data analysis tasks. The SCIKIT-LEARN framework will be introduced in more detail in Chap. 4.

References

1. Beyer, H., Schwefel, H.: Evolution strategies—a comprehensive introduction. Nat. Comput. **1**(1), 3–52 (2002)
2. Rechenberg, I.: Evolutionsstrategie—Optimierung technischer Systeme nach Prinzipien der biologischen Evolution. Frommann-Holzboog, Stuttgart (1973)
3. Schwefel, H.-P.: Numerische Optimierung von Computer-Modellen mittel der Evolutionsstrategie. Birkhaeuser, Basel (1977)
4. Holland, J.H.: Adaptation in Natural and Artificial Systems. University of Michigan Press, Ann Arbor (1975)
5. Kramer, O.: Evolution strategies with ledoit-wolf covariance matrix estimation. In: Proceedings of the IEEE Congress on Evolutionary Computation, CEC 2015, pp. 1712–1716 (2015)
6. Kramer, O.: Local fitness meta-models with nearest neighbor regression. In: Proceedings of the 19th European Conference on Applications of Evolutionary Computation, EvoApplications 2016, pp. 3–10. Porto, Portugal (2016)
7. Kramer, O.: Dimensionality reduction in continuous evolutionary optimization. In: 2015 International Joint Conference on Neural Networks, IJCNN 2015, pp. 1–4. Killarney, Ireland, 12–17 July 2015
8. Kramer, O., Lückehe, D.: Visualization of evolutionary runs with isometric mapping. In: Proceedings of the IEEE Congress on Evolutionary Computation, CEC 2015, pp. 1359–1363. Sendai, Japan, 25–28 May 2015
9. van der Walt, S., Colbert, S.C., Varoquaux, G.: The NumPy array: a structure for efficient numerical computation. Comput. Sci. Eng. **13**(2), 22–30 (2011)
10. Jones, E., Oliphant, T., Peterson, P., et al.: SciPy: open source scientific tools for Python (2001–2014). Accessed 03 Oct 2014
11. Hunter, J.D.: Matplotlib: a 2D graphics environment. Comput. Sci. Eng. **9**(3), 90–95 (2007)
12. Pedregosa, F., Varoquaux, G., Gramfort, A., Michel, V., Thirion, B., Grisel, O., Blondel, M., Prettenhofer, P., Weiss, R., Dubourg, V., Vanderplas, J., Passos, A., Cournapeau, D., Brucher, M., Perrot, M., Duchesnay, E.: Scikit-learn: machine learning in Python. J. Mach. Learn. Res. **12**, 2825–2830 (2011)

Part I
Evolution Strategies

Chapter 2
Evolution Strategies

2.1 Introduction

Many real-world problems have multiple local optima. Such problems are called multimodal optimization problems and are usually difficult to solve. Local search methods, i.e., methods that greedily improve solutions based on search in the neighborhood of a solution, often only find an arbitrary local optimum that may not be the global one. The most successful methods in global optimization are based on stochastic components, which allow escaping from local optima and overcome premature stagnation. A famous class of global optimization methods are ES. They are exceptionally successful in continuous solution spaces. ES belong to the most famous evolutionary methods for blackbox optimization, i.e., for optimization scenarios, where no functional expressions are explicitly given and no derivatives can be computed.

ES imitate the biological principle of evolution [1] and can serve as an excellent introduction to learning and optimization. They are based on three main mechanisms oriented to the process of Darwinian evolution, which led to the development of all species. Evolutionary concepts are translated into algorithmic operators, i.e., recombination, mutation, and selection.

First, we define an optimization problem formally. Let $f : \mathbb{R}^d \to \mathbb{R}$ be the fitness function to be minimized in the space of solutions $\mathbb{R}^d$. The problems we consider in this work are minimization problems unless explicitly stated, i.e., high fitness corresponds to low fitness function values. The task is to find a solution $\mathbf{x}^* \in \mathbb{R}^d$ such that $f(\mathbf{x}^*) \leq f(\mathbf{x})$ for all $\mathbf{x} \in \mathbb{R}^d$. A desirable property of an optimization method is to find the optimum $\mathbf{x}^*$ with fitness $f(\mathbf{x}^*)$ within a finite and preferably low number of function evaluations. Problem f can be an arbitrary optimization problem. However, we concentrate on continuous ones.

This chapter is structured as follows. Section 2.2 gives short introduction to the basic principles of EAs. The history of evolutionary computation is sketched in Sect. 2.3. The evolutionary operators are presented in the following sections, i.e., recombination in Sect. 2.4, mutation in Sect. 2.5, and selection in Sect. 2.6,

© Springer International Publishing Switzerland 2016
O. Kramer, *Machine Learning for Evolution Strategies*,
Studies in Big Data 20, DOI 10.1007/978-3-319-33383-0_2

respectively. Step size control is an essential part of the success of EAs and is introduced in Sect. 2.7 with Rechenberg's rule. As the (1+1)-ES has an important part to play in this book, Sect. 2.8 is dedicated to this algorithmic variant. The chapter closes with conclusions in Sect. 2.9.

2.2 Evolutionary Algorithms

If derivatives are available, Newton methods and variants are the proper algorithmic choices. From this class of methods, the Broyden-Fletcher-Goldfarb-Shanno (BFGS) algorithm [2] belongs to the state-of-the-art techniques in optimization. In this book, we concentrate on blackbox optimization problems. In blackbox optimization, the problem does not have to fulfill any assumptions or limiting properties. For such general optimization scenarios, evolutionary methods are a good choice. EAs belong to the class of stochastic derivative-free optimization methods. Their biological motivation has made them very popular. They are based on recombination, mutation, and selection. After decades of research, a long history of applications and theoretical investigations have proven the success of evolutionary optimization algorithms.

Algorithm 1 EA

1: initialize $\mathbf{x}_1, \ldots, \mathbf{x}_\mu$

2: **repeat**

3:    **for** $i = 1$ **to** λ **do**

4:       select ρ parents

5:       recombination $\rightarrow \mathbf{x}_i'$

6:       mutate $\mathbf{x}_i'$

7:       evaluate $\mathbf{x}_i' \rightarrow f(\mathbf{x}_i')$

8:    **end for**

9:    select μ parents from $\{\mathbf{x}_i'\}_{i=1}^\lambda \rightarrow \{\mathbf{x}_i\}_{i=1}^\mu$

10: **until** termination condition

In the following, we shortly sketch the basic principles of EAs oriented to Algorithm 1. Evolutionary search is based on a set $\{\mathbf{x}_1, \ldots, \mathbf{x}_\mu\}$ of parental and a set $\{\mathbf{x}_1, \ldots, \mathbf{x}_\lambda\}$ of offspring candidate solutions, also called individuals. The solutions are iteratively subject to changes and selection of the best offspring candidates. In the generational loop, λ offspring solutions are generated. For each offspring solution, the recombination operator selects ρ parents and combines their parts to a new candidate solution. The mutation operator adds random changes, i.e. noise, to the preliminary candidate resulting in solution $\mathbf{x}_i'$. Its quality in solving the optimization problem is called fitness and is evaluated on fitness function $f(\mathbf{x}_i')$. All candidate solutions of a generation are put into offspring population $\{\mathbf{x}_i'\}_{i=1}^\lambda$. At the end of a generation, μ solutions are selected and constitute the novel parental population $\{\mathbf{x}_i\}_{i=1}^{\mu}$ that is basis of the following generation.

The optimization process is repeated until a termination condition is reached. Typical termination conditions are defined via fitness values or via an upper bound on the number of generations. In the following, we will shortly present the history of evolutionary computation, introduce evolutionary operators, and illustrate concepts that have proven well in ES.

2.3 History

In the early 1950s, the idea came up to use algorithms for problem solving that are oriented to the concept of evolution. In Germany, the history of evolutionary computation began with ES, which were developed by Rechenberg and Schwefel in the sixties and seventies of the last century in Berlin [3–5]. At the same time, Holland introduced the evolutionary computation concept in the United States known as genetic algorithms [6]. Also Fogel introduced the idea at that time and called this approach evolutionary programming [7]. For about 15 years, the disciplines developed independently from each other before growing together in the 1980s. Another famous branch of evolutionary computation was proposed in the nineties of the last century, i.e., genetic programming (GP) [8]. GP is about evolving programs by means of evolution. These programs can be based on numerous programming concepts and languages, e.g., assembler programs or data structures like trees. Genetic programming operators are oriented to similar principles like other EAs, but adapted to evolving programs. For example, recombination combines elements of two or more programs. In tree representations, subtrees are exchanged. Mutation changes a program. In assembler code, a new command may be chosen. In tree representations, a new subtree can be generated. Mutation can also lengthen or shorten programs.

Advanced mutation operators, step size mechanisms, and methods to adapt the covariance matrix like the CMA-ES [9] have made ES one of the most successful optimizers in derivative-free continuous optimization. For binary, discrete, and combinatorial representations, other concepts are known. Annual international conferences like the Genetic and Evolutionary Computation Conference (GECCO), the Congress on Evolutionary Computation (CEC), and EvoStar in Europe contribute to the understanding and distribution of EAs as solid concepts and search methods.

Related to evolutionary search are estimation of distribution algorithms (EDAs) and particle swarm optimization (PSO) algorithms. Both are based on randomized operators like EAs, while PSO algorithms are also nature-inspired. PSO models the flight of solutions in the solution space with velocities, while being oriented to the best particle positions. All nature-inspired methods belong to the discipline computational intelligence, which also comprises neural networks and fuzzy-logic. Neural networks are inspired by natural neural processing, while fuzzy logic is a logic inspired by the fuzzy way of human language and concepts.

2.4   Recombination

Recombination, also known as crossover, mixes the genetic material of parents. Most evolutionary algorithms also make use of a recombination operator that combines the information of two or more candidate solutions $\mathbf{x}_1, \ldots, \mathbf{x}_\rho$ to a new offspring solution. Hence, the offspring carries parts of the genetic material of its parents. Many recombination operators are restricted to two parents, but also multi-parent recombination variants have been proposed in the past that combine information of ρ parents. The use of recombination is discussed controversially within the building block hypothesis by Goldberg [10], Holland [11]. The building block hypothesis assumes that good solution substrings called building blocks are spread over the population in the course of the evolutionary process, while their number increases.

For bit strings and similar representations, multi-point crossover is a common recombination operator. It splits up the representations of two ore more parents at multiple positions and combines the parts alternately to a new solution.

Typical recombination operators for continuous representations are dominant and intermediate recombination. Dominant recombination randomly combines the genes of all parents. With ρ parents $\mathbf{x}_1, \ldots, \mathbf{x}_\rho \in \mathbb{R}^d$, it creates the offspring solution $\mathbf{x}' = (x'_1, \ldots, x'_d)^T$ by randomly choosing the i-th component

$$x'_i = (x_i)_j, \quad j \in \text{ random } \{1, \ldots, \rho\}. \tag{2.1}$$

Intermediate recombination is appropriate for numerical solution spaces. Given ρ parents $\mathbf{x}_1, \ldots, \mathbf{x}_\rho$ each component of the offspring vector $\mathbf{x}'$ is the arithmetic mean of the components of all ρ parents

$$x'_i = \frac{1}{\rho} \sum_{j=1}^{\rho} (x_i)_j. \tag{2.2}$$

The characteristics of offspring solutions lie between their parents. Integer representations may require rounding procedures for generating valid solutions.

2.5   Mutation

Mutation is the second main source of evolutionary changes. The idea of mutation is to add randomness to the solution. According to Beyer and Schwefel [3], a mutation operator is supposed to fulfill three conditions. First, from each point in the solution space each other point must be reachable. This condition shall guarantee that the optimum can be reached in the course of the optimization run. Second, in unconstrained solution spaces a bias is disadvantageous, because the direction to the optimum is unknown. By avoiding a bias, all directions in the solution space can be reached with the same probability. This condition is often hurt in practical optimization to accelerate the search. Third, the mutation strength should be adjustable, in order to adapt exploration and exploitation to local solution space conditions, e.g., to accelerate the search when the success probability is high.

For bit string representations, the bit-flip mutation operator is usual. It flips each bit, i.e., changes a 0 to 1 and a 1 to 0 with probability $1/d$, if d is the number of bits. In $\mathbb{R}^d$ the Gaussian mutation operator is common. Let $\mathcal{N}(0, 1)$ represent a randomly drawn Gaussian distributed number with expectation 0 and standard deviation 1. Mutation adds Gaussian noise to each solution candidate using

$$\mathbf{x}' = \mathbf{x} + \mathbf{z}, \tag{2.3}$$

with a mutation vector $\mathbf{z} \in \mathbb{R}^d$ based on sampling

$$\mathbf{z} \sim \sigma \cdot \mathcal{N}(0, 1). \tag{2.4}$$

The standard deviation σ plays the role of the mutation strength and is also known as step size. The isotropic Gaussian mutation with only one step size uses the same standard deviation for each component x_i. The convergence towards the optimum can be improved by adapting σ according to local solution space characteristics. In case of high success rates, i.e., a large number of offspring solutions being better than their parents, big step sizes are advantageous, in order to explore the solution space as fast as possible. This is often reasonable at the beginning of the search. In case of low success rates, smaller step sizes are appropriate. This is often adequate in later phases of the search during convergence to the optimum, i.e., when good evolved solutions should not be destroyed. An example for an adaptive control of step sizes is the 1/5-th success rule by Rechenberg [4] that increases the step size, if the success rate is over 1/5-th, and decreases it, if the success rate is lower. The Rechenberg rule will be introduced in more detail in Sect. 2.7.

2.6 Selection

The counterpart of the variation operators mutation and recombination is selection. ES usually do not employ a competitive selection operator for mating selection. Instead, parental solutions are randomly drawn from the set of candidate solutions. But for survivor selection, the elitist selection strategies comma and plus are used. They choose the μ-best solutions as basis for the parental population of the following generation. Both operators, plus and comma selection, can easily be implemented by sorting the population with respect to the solutions' fitness. Plus selection selects the μ-best solutions from the union $\{\mathbf{x}_i\}_{i=1}^{\mu} \cup \{\mathbf{x}_i'\}_{i=1}^{\lambda}$ of the last parental population $\{\mathbf{x}_i\}_{i=1}^{\mu}$ and the current offspring population $\{\mathbf{x}_i'\}_{i=1}^{\lambda}$, and is denoted as $(\mu + \lambda)$-ES. In contrast, comma selection, i.e. (μ, λ)-ES, selects exclusively from the offspring population, neglecting the parental population, even if the parents have a superior fitness. Forgetting superior solutions may sound irrational. But good solutions can be only local optima. The evolutionary process may fail to leave them without the ability to forget.

2.7 Rechenberg's 1/5th Success Rule

The adjustment of parameters and adaptive operator features is of crucial importance for reliable results and the efficiency of evolutionary heuristics. Furthermore, proper parameter settings are important for the comparison of different algorithms. The problem arises how evolutionary parameters can be tuned and controlled. The change of evolutionary parameters during the run is called online parameter control and is reasonable, if the conditions of the fitness landscape change during the optimization process. In deterministic control, parameters are adjusted according to a fixed time scheme, e.g., depending on the generation number like proposed by Fogarty [12] and by Bäck and Schütz [13]. However, it may be useful to reduce the mutation strengths during the evolutionary search, in order to allow convergence. For example, running a (1+1)-ES with isotropic Gaussian mutations with constant step sizes σ, the optimization process will become slow after a certain number of generations. A deterministic scheme may fail to hit the optimal speed the step sizes are reduced and may also not be able to increase them if necessary.

The step sizes have to be adapted in order to speed up the optimization process in a more flexible way. A powerful scheme is the 1/5th success rule by Rechenberg [4], which adapts step size σ for optimal progress. In case of the (1+1)-ES, the optimization process runs for a number T of generations. During this period, step size σ is kept constant and the number T_s of successful generations is counted. From T_s, the success probability p_s is estimated by

$$p_s = T_s/T \tag{2.5}$$

and step size σ is changed according to

$$\sigma = \begin{cases} \sigma/\tau, & \text{if } p_s > 1/5 \\ \sigma \cdot \tau, & \text{if } p_s < 1/5 \\ \sigma, & \text{if } p_s = 1/5 \end{cases}$$

with $0 < \tau < 1$. This control of the step size is implemented in order adapt to local solution space characteristics and to speed up the optimization process in case of large success probabilities. Figure 2.1 illustrates the Rechenberg rule with $T = 5$. The fitness is increasing from left to right. The blue candidate solutions are the successful solutions of a (1+1)-ES, the grey ones are discarded due to worse fitness. Seven mutation steps are shown, at the beginning with a small step size, illustrated by the smaller light circles. After five mutations, the step size is increased, as the success rate is larger than 1/5, i.e. $p_s = 3/5$, illustrated by larger dark circles. Bigger steps

Fig. 2.1 Illustration of step size adaptation with Rechenberg's success rule. Three of five steps are successful resulting in a success probability of $p_s = 3/5$ and an increase of step size σ

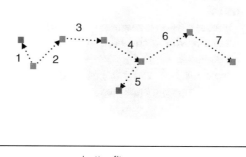

better fitness

into the direction of the optimum are possible. The objective of Rechenberg's step size adaptation is to stay in the evolution window guaranteeing optimal progress. The optimal value for τ depends on various factors such as the number T of generations and the dimension d of the problem.

A further successful concept for step size adaptation is self-adaptation, i.e., the automatic evolution of the mutation strengths. In self-adaptation, each candidate is equipped with an own step size, which is subject to recombination and mutation. Then, the objective variable is mutated with the inherited and modified step size. As solutions consist of objective variables and step sizes, the successful ones are selected as parents for the following generation. The successful step sizes are spread over the population.

2.8 (1+1)-ES

The (1+1)-ES with Gaussian mutation and Rechenberg's step size control is the basis of the evolutionary algorithms used in this book. We choose this method to reduce side effects. The more complex algorithms are, the more probable are side effects changing the interactions with machine learning extensions. We concentrate on the (1+1)-ES, which is well understood from a theoretical perspective. Algorithm 2 shows the pseudocode of the (1+1)-ES. After initialization of $\mathbf{x}$ in $\mathbb{R}^d$, the evolutionary loop begins. The solution $\mathbf{x}$ is mutated to $\mathbf{x}'$ with Gaussian mutation and step size σ that is adapted with Rechenberg's 1/5th rule. The new solution $\mathbf{x}'$ is accepted, if its fitness is better than or equal to the fitness of its parent $\mathbf{x}$, i.e., if $f(\mathbf{x}') \leq f(\mathbf{x})$. Accepting the solution in case of equal fitness is reasonable to allow random walk on plateaus, which are regions in solution space with equal fitness. For the (1+1)-ES and all variants used in the remainder of this book, the fitness for each new solution $\mathbf{x}'$ is computed only once, although the condition in Line 6 might suggest another fitness function call is invoked.

Algorithm 2 (1+1)-ES

1: intialize $\mathbf{x}$
2: **repeat**
3:    mutate $\mathbf{x}' = \mathbf{x} + \mathbf{z}$ with $\mathbf{z} \sim \sigma \cdot \mathcal{N}(0, 1)$
4:    adapt σ with Rechenberg
5:    evaluate $\mathbf{x}' \to f(\mathbf{x}')$
6:    replace $\mathbf{x}$ with $\mathbf{x}'$ if $f(\mathbf{x}') \leq f(\mathbf{x})$
7: **until** termination condition

The evolutionary loop is repeated until a termination condition is reached. The (1+1)-ES shares similarities with simulated annealing, but employs Gaussian mutation with step size adaptation, while not using the cooling scheme for accepting a worse solution. For multimodal optimization, restarts may be required as the diversity of the (1+1)-ES without a population is restricted to only one single solution. The (1+1)-ES is the basic algorithm for the optimization approaches introduced in Chaps. 3, 4, 7, 9, and 10.

2.9   Conclusions

Evolutionary algorithms are famous blackbox optimization algorithms. They are based on a population of candidate solutions or one single solution in case of the (1+1)-ES. Evolutionary optimization algorithms are inspired by the idea of natural selection and Darwinian evolution. They have their roots in the fifties and sixties of the last century. Meanwhile, ES have developed to outstandingly successful blackbox optimization methods for continuous solution spaces. For exploration of the solution space, recombination operators combine the properties of two or more solutions. Mutation operators use noise to explore the solution space. Selection exploits the search by choosing the best solutions to be parents for the following generation. Parameter control techniques like Rechenberg's 1/5th success rule improve the convergence towards the optimum. In case of success, larger steps can be taken in solution space, while in case of stagnation, progress is more probable when the search concentrates on the close environment of the current solution.

Extensions allow ES to search in constrained or multi-objective solution spaces. In multi-objective optimization, the task is to evolve a Pareto set of non-dominated solutions. Various strategies are available for this sake. An example is NSGA-ii [14] that maximizes the Manhattan distance of solutions in objective space to achieve a broad coverage on the Pareto front. Theoretical results are available for discrete and continuous algorithmic variants and solution spaces. An example for a simple result is the runtime of a (1+1)-EA with bit flip mutation on the function OneMax that maximizes the number of ones in a bit string. The expected runtime is upper bound by $O(d \log d)$, if d is the length of the bit string [15]. The idea of the proof is based on the lemma for fitness-based partitions, which divides the solution space into disjoint sets of solutions with equal fitness and makes assertions about the expected time required to leave these partitions.

This book will concentrate on extensions of ES with machine learning methods to accelerate and support the search. The basic mechanisms of ES will be extended by covariance matrix estimation, fitness function surrogates, constraint function surrogates, and dimensionality reduction approaches for optimization, visualization, and niching.

References

1. Kramer, O., Ciaurri, D.E., Koziel, S.: Derivative-free optimization. In: Computational Optimization and Applications in Engineering and Industry. Springer (2011)
2. Nocedal, J., Wright, S.J.: Numerical Optimization. Springer (2000)
3. Beyer, H., Schwefel, H.: Evolution strategies—A comprehensive introduction. Nat. Comput. **1**(1), 3–52 (2002)
4. Rechenberg, I.: Evolutionsstrategie—Optimierung technischer Systeme nach Prinzipien der biologischen Evolution. Frommann-Holzboog, Stuttgart (1973)
5. Schwefel, H.-P.: Numerische Optimierung von Computer-Modellen mittel der Evolutionsstrategie. Birkhaeuser, Basel (1977)
6. Holland, J.H.: Adaptation in Natural and Artificial Systems. University of Michigan Press, Ann Arbor (1975)
7. Fogel, D.B.: Evolving artificial intelligence. PhD thesis, University of California, San Diego (1992)
8. Koza, J.R.: Genetic Programming: On the Programming of Computers by Means of Natural Selection. MIT Press, Cambridge (1992)
9. Hansen, N., Ostermeier, A.: Adapting arbitrary normal mutation distributions in evolution strategies: the covariance matrix adaptation. In: International Conference on Evolutionary Computation, pp. 312–317 (1996)
10. Goldberg, D.: Genetic Algorithms in Search. Optimization and Machine Learning. Addison-Wesley, Reading, MA (1989)
11. Holland, J.H.: Hidden Order: How Adaptation Builds Complexity. Addison-Wesley, Reading, MA (1995)
12. Fogarty, T.C.: Varying the probability of mutation in the genetic algorithm. In: Proceedings of the 3rd International Conference on Genetic Algorithms, pp. 104–109. Morgan Kaufmann Publishers Inc, San Francisco (1989)
13. Bäck, T., Schüz, M.: Intelligent mutation rate control in canonical genetic algorithms. In: Proceedings of the 9th International Symposium on Foundation of Intelligent Systems, ISMIS 1996, pp. 158–167. Springer (1996)
14. Deb, K., Agrawal, S., Pratap, A., Meyarivan, T.: A fast elitist non-dominated sorting genetic algorithm for multi-objective optimisation: NSGA-II. In: Proceedings of the 6th International Conference on Parallel Problem Solving from Nature, PPSN VI 2000, pp. 849–858. Paris, France, 18–20 Sept 2000
15. Droste, S., Jansen, T., Wegener, I.: On the analysis of the (1+1) evolutionary algorithm. Theoret. Comput. Sci. **276**(1–2), 51–81 (2002)

Chapter 3
Covariance Matrix Estimation

3.1 Introduction

Covariance matrix estimation allows the adaptation of Gaussian-based mutation operators to local solution space characteristics. The covariance matrix adaptation evolution strategy (CMA-ES) [1] is an example for a highly developed ES using covariance matrix estimation. Besides the cumulative path length control, it iteratively adapts the covariance matrix for the Gaussian mutation operator. In numerous artificial benchmark analyses and real-world applications, covariance matrix estimation has proven its strengths. But it is not restricted to the adaptive update mechanism of the CMA-ES. Efficient implementations allow the fast estimation of the covariance matrix with specialized methods based on the current evolutionary population or based on a training set of successful solutions.

In this chapter, we propose a covariance matrix estimation variant using a (1+1)-ES with Gaussian mutation and Rechenberg's adaptive step size control rule. The idea is to estimate the covariance matrix of the last N successful solutions corresponding to successful generations and to sample from the Gaussian distribution employing the estimated covariance matrix. For covariance matrix estimation, we employ maximum likelihood-based empirical covariance matrix estimation and the Ledoit-Wolf method [2]. The latter shrinks the empirical covariance matrix and is know to be a powerful extension in practice.

This chapter is structured as follows. Section 3.2 gives a brief introduction to covariance matrix estimation focusing on empirical estimation and the Ledoit-Wolf estimator, while Sect. 3.3 presents the covariance matrix estimation ES, which we abbreviate with COV-ES in the following. Related work is discussed in Sect. 3.4. The COV-ES is experimentally analyzed in Sect. 3.5. Conclusions are drawn in Sect. 3.6.

© Springer International Publishing Switzerland 2016
O. Kramer, *Machine Learning for Evolution Strategies*,
Studies in Big Data 20, DOI 10.1007/978-3-319-33383-0_3

3.2 Covariance Matrix Estimation

Covariance matrix estimation is the task to find the unconstrained and statistically interpretable parameters of a covariance matrix. It is still an actively investigated research problem in statistics. Given a set of points $\mathbf{x}_1, \ldots, \mathbf{x}_N \in \mathbb{R}^d$, the task is to find the covariance matrix $\mathbf{C} \in \mathbb{R}^{d \times d}$. The estimation of $\mathbf{C}$ has an important part to play in various fields like time series analysis, classical multivariate statistics and data mining. A common approach for estimating $\mathbf{C}$ is the maximum likelihood approach. The log-likelihood function is maximized by the sample covariance

$$\mathbf{S} = \frac{1}{N} \sum_{i=1}^{N} \mathbf{x}_i \mathbf{x}_i^T. \tag{3.1}$$

This sample covariance is an unbiased estimator of the population covariance matrix and requires a sufficiently large data set. For covariance matrix estimation in an ES this means that the training set size N has to be large enough for a robust covariance matrix estimation. But as the distribution changes during the optimization run, in particular getting narrower when approximating the optimum, N also has an upper limit. Due to the curse of dimensionality problem the requirements on the populations size grow enormously with increasing problem dimension d. Probability distributions and the maximum likelihood estimation are introduced in detail by Bishop [3].

As the maximum likelihood method is not a good estimator of the eigenvalues of the covariance matrix, the shrinkage method has been introduced. In the Ledoit-Wolf covariance matrix estimation [2], the covariance matrix estimator obtained from Sharpe's single-index model [4] joins the sample covariance matrix in a weighted average. Assume, we have given the sample covariance matrix $\mathbf{S}$. Idea of the shrinkage methods is to compute a weighted covariance matrix from $\mathbf{S}$ and a shrinked but more structured matrix $\mathbf{F}$, which is the shrinkage target

$$\mathbf{C} = \phi \cdot \mathbf{S} + (1 - \phi) \cdot \mathbf{F} \tag{3.2}$$

with weight vector ϕ. $\mathbf{F}$ is also known as sample constant correlation matrix. It is defined via sample variances and average sample correlations, see [2]. The question comes up for the optimal weight ϕ^* between both matrices. For this sake, a loss is defined that measures the deviation of the shrinkage estimator $\mathbf{C}$ and the true covariance matrix Σ

$$L(\phi) = \|\mathbf{C} - \Sigma\|_F^2 \tag{3.3}$$

with Frobenius norm $\| \cdot \|_F^2$. The estimation risk is the corresponding expected value of the loss. From this loss the optimal value ϕ^* is proposed to be chosen as

$$\phi^* = \max \left\{ 0, \min \left\{ \frac{\kappa}{T}, 1 \right\} \right\} \tag{3.4}$$

with parameter κ. For a detailed derivation see [2]. We employ the covariance matrix estimators from the SCIKIT- LEARN library.

- The command `from sklearn.covariance import LedoitWolf` imports the Ledoit-Wolf covariance matrix estimator.
- Similarly, `from sklearn.covariance import EmpiricalCovariance` imports the empirical covariance matrix estimator for comparison.
- `LedoitWolf().fit(X)` trains the Ledoit-Wolf estimator with set X of patterns. The estimator saves the corresponding covariance matrix in attribute `covariance_`.
- `numpy.linalg.cholesky(C)` computes the Cholesky decomposition of C. The result is multiplied with a random Gaussian vector scaled by step size `sigma` with `numpy.dot(C_,sigma * np.random.randn(N))`.

3.3 Algorithm

The integration of the covariance matrix estimation into the ES optimization framework is described in the following. We employ a (1+1)-ES with Rechenberg rule for step size control. The population for the estimation process is based on the history of the best solutions that have been produced in the course of the optimization process.

Algorithm 3 shows the pseudocode of the covariance matrix estimation-based ES with Ledoit-Wolf covariance estimation. In the first phase, the start solution $\mathbf{x}$ is initialized, while the covariance matrix is initialized with the identity matrix $\mathbf{C} = \mathbf{I}$. In the main generational loop, an offspring solution $\mathbf{x}'$ is generated with

$$\mathbf{x}' = \mathbf{x} + \mathbf{z} \tag{3.5}$$

based on the Gaussian distribution using the Cholesky decomposition for computing $\sqrt{\mathbf{C}}$ of covariance matrix $\mathbf{C}$ for employing

$$\mathbf{z} \sim \sigma \cdot \sqrt{\mathbf{C}} \mathcal{N}(\mathbf{0}, \mathbf{I}). \tag{3.6}$$

Step size σ is adapted with Rechenberg's 1/5th success rule. The offspring solution $\mathbf{x}'$ is accepted, if its fitness is superior, i.e., if $f(\mathbf{x}') \leq f(\mathbf{x})$. From the set of the last N successful solutions $\{\mathbf{x}_i\}_{i=1}^N$, a new covariance matrix $\mathbf{C}$ is estimated. This training set of successful solutions forms the basis of the covariance matrix estimation population. Novel mutations will be sampled from this estimated covariance matrix and will consequently be similar to the successful past solutions, which allows the local approximation of the solution space and a movement towards promising regions. For the covariance matrix estimation, we use the empirical estimation and the Ledoit-Wolf approach introduced in the previous section. These steps are repeated until a termination condition is fulfilled.

Algorithm 3 COV-ES

1: initialize $\mathbf{x}$
2: $\mathbf{C} = \mathbf{I}$
3: **repeat**
4:    adapt σ with Rechenberg
5:    $\mathbf{z} \sim \sigma \cdot \sqrt{\mathbf{C}} \mathcal{N}(\mathbf{0}, \mathbf{I})$
6:    $\mathbf{x}' = \mathbf{x} + \mathbf{z}$
7:    evaluate $\mathbf{x}' \to f(\mathbf{x}')$
8:    **if** $f(\mathbf{x}') \le f(\mathbf{x})$ **then**
9:       replace $\mathbf{x}$ with $\mathbf{x}'$
10:      last N solutions $\to \{\mathbf{x}_i\}_{i=1}^{N}$
11:      covariance estimation $\{\mathbf{x}_i\}_{i=1}^{N} \to \mathbf{C}$
12:   **end if**
13: **until** termination condition

3.4 Related Work

The CMA-ES by Hansen and Ostermeier [1] is a successful evolutionary optimization method for real-parameter optimization of non-linear problems. It is successfully applied to various domains, e.g., to solving satisfiability in fuzzy logics [5], for tuning prediction models that serve as surrogates of expensive finite-element simulations [6], and to search for good degree distributions with different decoding behavior in luby transform code [7].

The covariance update in each step of the CMA-ES is similar to an approximation of the inverse Hessian matrix and requires a set of different equations [1]. Moreover, the approach is based on a derandomized step size control that takes into account how solutions move in the course of the optimization process. To avoid the computationally expensive complexity of $O(N^3)$ for the Cholesky decomposition in each iteration, Igel et al. [8] propose an alternative covariance matrix update equation. An efficient update rule that allows an update in $O(N^2)$ is proposed by Suttorp et al. [9], which also removes the need for matrix inversion. The CMA-ES is continuously improved. A computationally efficient variant for limited memory is introduced in [10], a variant capable of handling noisy objective functions has been introduced by Kruisselbrink et al. [6]. The CMA-ES is combined with local search by Caraffini et al. [11]. CMA-ES variants are proposed for multi-objective optimization [12–14] and constraint handling [15]. To avoid stagnation in local optima, the CMA-ES can be combined with restarts, which is experimentally analyzed on the CEC 2013 benchmark problems in [16]. A CMA-ES variant based on self-adaptation is the CMSA-ES by Beyer and Sendhoff [17]. It applies self-adaptive step size control, i.e., step sizes are inherited with the solutions they generated. The CMSA-ES employs the usual empirical covariance matrix update. Au and Leung [18] propose to cluster the eigenvalues of the covariance matrix of a CMA-ES and to sample search points on a mirrored eigenspace spanned by eigenvectors. Krause and Glasmachers [19] propose a multiplicative update rule for the CMA-ES.

EDAs are similar to the COV-ES. As population-based approaches they estimate the distribution of the best solutions assuming a certain shape of the population distribution. An EDA variant based on Ledoit-Wolf estimation is the Shrinkage Estimation of Distribution Algorithm for the Multivariate Norm (SEDA) by Ochoa [20]. It works like a classic EDA without step size adaptation and ES-typical elements.

3.5 Experimental Analysis

In this section, we experimentally analyze the COV-ES by visualizing the estimation during evolutionary runs and by comparing the performance on a short benchmark test set. In the optimal case, the covariance matrix estimation approximates the local contour lines of the fitness function. This allows sampling new mutations according to the local fitness function conditions. Figure 3.1 shows visualizations of the Ledoit-Wolf estimation of the best solutions generated during the run of the ES on the Sphere function and on Rosenbrock. After 200 generations of the (1+1)-ES, the covariance matrix estimation based on the last $N = 100$ best solutions is illustrated. The plots visualize the contour lines of the fitness functions and the contour lines of the covariance matrix estimate. In particular on the Sphere function, the covariance matches the contour lines of the fitness function quite well. On Rosenbrock, we can observe that the search is turning around the curved landscape towards the optimum.

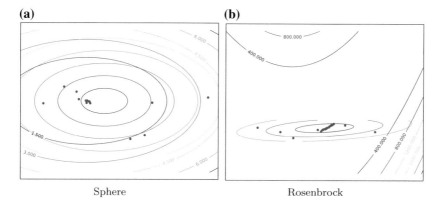

(a)   **(b)**

Sphere   Rosenbrock

Fig. 3.1 Covariance matrix estimation of last $N = 100$ solutions of a (1+1)-ES after 200 generations. **a** On the Sphere function, Ledoit-Wolf allows a good adaptation of the covariance to the fitness contour lines. **b** On the Rosenbrock function, the covariance is adapting to the curved landscape towards the optimum

In the following, we compare the variants to the standard (1+1)-ES without covariance matrix estimation, i.e., with $\mathbf{C} = \mathbf{I}$ during optimization runs. The Rechenberg rule employs $T = 10$ and $\tau = 0.5$. For the covariance matrix estimation process, a population size of $N = 100$ is used. The ES terminate after 5000 fitness function evaluations. Table 3.1 shows the experimental analysis on the two functions Sphere and Rosenbrock with $d = 5$ and $d = 10$ dimensions. Each experiment is repeated 100 times and the mean values and corresponding standard deviations are shown. The last columns show the p-value of a Wilcoxon test [21] comparing the empirical covariance matrix estimation with the Ledoit-Wolf version.

The results show that the COV-ES outperforms the standard (1+1)-ES without covariance matrix estimation on Rosenbrock in both dimensions. While the Ledoit-Wolf estimation is able to adapt to the narrow valley when approximating the optimum, the standard (1+1)-ES fails. For $d = 5$, the empirical covariance matrix estimation ES variant is better than the classic (1+1)-ES, but is outperformed by the Ledoit-Wolf variant. For $d = 10$, the advantage of the covariance matrix estimation mechanism becomes even more obvious. The low p-values of the Wilcoxon test confirm the statistical significance of the result. On the Sphere function, the (1+1)-ES is slightly, but not significantly superior for $d = 5$, but shows significantly better results than both estimation variants for $d = 10$ (p-value 0.047). This is due to the fact that isotropic Gaussian mutation is the optimal setting on the symmetric curvature of the Sphere. The empirical covariance matrix estimation fails for $d = 10$, where Ledoit-Wolf is significantly superior. The ES with Ledoit-Wolf performs significantly worse on the Cigar with $d = 5$, but better for $d = 10$ dimensions, while no statistical difference can be observed on Griewank.

Figures 3.2 and 3.3 show comparisons of evolutionary runs of the ES with empirical covariance estimation and of the COV-ES on the four benchmark functions with $d = 10$ and a logarithmic scale of the fitness. Again, the ES terminate after 5000 fitness function evaluations. The plots show the mean evolutionary runs, while the

Table 3.1 Experimental comparison between (1+1)-ES and both COV-ES variants. Bold values indicate statistical significance with p-value of Wilcoxon test < 0.05

Problem		(1+1)-ES		Empirical		Ledoit-Wolf		Wilx.
	d	Mean	Dev	Mean	Dev	Mean	Dev	p-Value
Sphere	5	5.91e-31	6.54e-31	3.35e-30	2.97e-30	3.74e-30	3.54e-30	0.197
	10	7.97e-30	8.89e-30	0.177	0.237	4.33e-24	7.50e-24	**0.047**
Rosenbrock	5	0.286	0.278	6.88e-06	1.13e-05	1.69e-23	2.926-23	0.144
	10	0.254	0.194	5.138	4.070	1.23e-17	6.99e-18	**0.007**
Cigar	5	6.855	6.532	0.031	0.080	16.030	18.825	**0.007**
	10	7.780	11.132	2.35e5	4.78e5	17.544	28.433	**0.007**
Griewank	5	0.012	0.008	0.018	0.018	0.012	0.009	0.260
	10	0.010	0.010	0.020	0.014	0.008	0.005	0.144

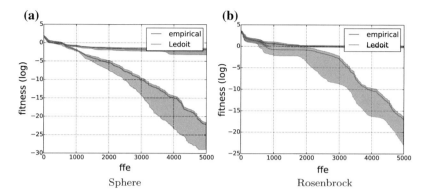

Fig. 3.2 Comparison of empirical covariance matrix estimation and Ledoit-Wolf estimation (**a**) on the Sphere function. The COV-ES allows a logarithmically linear approximation of the optimum. **b** Also on Rosenbrock, the COV-ES allows a logarithmically linear approximation of the optimum

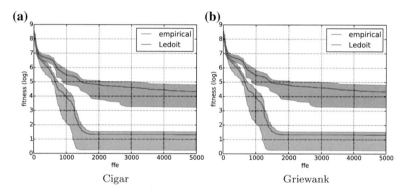

Fig. 3.3 **a** Although the smooth approximation of the optimum fails on the Cigar function, Ledoit-Wolf outperforms the empirical covariance matrix estimation. **b** Also on Griewank, the COV-ES outperforms the empirical covariance matrix estimation

upper and lower parts illustrate the best and worst runs. All other runs lie in the shadowed regions. The figures show that the Ledoit-Wolf variant is superior to the empirical variant on all four problems. Ledoit-Wolf allows a logarithmically linear development on most of the functions, particularly on the Sphere function and Rosenbrock. On Cigar and Griewank, the ES stagnates before reaching the optimum, which has already been shown in Table 3.1.

Last, we analyze the behavior of the CMA-ES with Ledoit-Wolf estimation w.r.t. various covariance matrix estimation training set sizes N and an increasing problem dimensionality on the Sphere function. The experimental results of the COV-ES with corresponding settings are shown in Table 3.2, where the mean fitness values and the standard deviations of 100 runs are shown with a termination of each run after 5000 fitness function evaluations. The best results are marked in bold.

Table 3.2 Analysis of covariance matrix estimation size N on fitness on Sphere function (Sp.) w.r.t. an increasing problem dimensionality d, and on Rosenbrock (Ro.) with $d = 10$

problem	$N = 20$	$N = 50$	$N = 100$
Sp. d=10	**2.83e-35** $\pm$ 9.1e-35	5.40e-29 $\pm$ 1.7e-28	2.30e-24 $\pm$ 5.6e-24
Sp. d=20	**1.39e-13** $\pm$ 1.6e-13	4.54e-09 $\pm$ 6.6e-09	1.41e-06 $\pm$ 2.5e-06
Sp. d=50	**5.59e-04** $\pm$ 1.1e-04	5.09e-02 $\pm$ 2.4e-02	5.49e-01 $\pm$ 2.6e-01
Ro. d=10	1.0017 $\pm$ 1.72	4.21e-08 $\pm$ 7.3e-08	**4.92e-18** $\pm$ 8.4e-18

We can observe that for all problem dimensions d, the best choice on the Sphere function is a low training set size N. There is a lower limit for the choice of N, i.e., values under $N = 20$ can lead to numerical problems. A larger N obviously slows down the optimization speed. This is probably due to the fact that older solutions are not appropriate to allow an estimate of the covariance matrix that can be exploited for good new mutations. The distributions become narrower when approximating the optimum. As expected, the optimization problem becomes harder for larger problem dimensions. The effect of perturbation of the optimization is weakened for large d.

3.6 Conclusions

In this chapter, we introduce an ES based on Ledoit-Wolf covariance matrix estimation. Due to changing distributions during evolutionary optimization processes, an adaptation of the probability distribution for the mutation operator is reasonable. In the exploration phase, the search is broadly scattered in solution space. During convergence phases, the search gets narrower. For Gaussian mutation, a covariance matrix can be employed that flexibly adapts to changing solution space conditions.

Due to efficient and easy-to-use implementations in machine learning libraries, there is no reason to employ alternative update rules that are more efficient to compute but less accurate. The Ledoit-Wolf method outperforms empirical covariance matrix estimation on the tested benchmark problems. Further, it allows smaller population sizes on high-dimensional solution spaces and thus a faster optimization process. In the ES we introduce in this Chapter, we employ the covariance matrix estimation process for a (1+1)-ES managing a training set of the last best solutions. The experimental results show that the evolutionary search can be accelerated significantly. Further, a small archive size turns out to be advantageous, probably due to the argument of changing distributions during the search.

The covariance matrix estimation with Ledoit-Wolf estimation allows a fast and easy modification of ES for the adaptation to certain fitness landscapes and solution spaces. The application of covariance matrix estimation to population-based ES is a straightforward undertaking. In each generation, the new parental population consisting of μ parents is subject to the covariance matrix estimation process.

References

1. Hansen, N., Ostermeier, A.: Adapting arbitrary normal mutation distributions in evolution strategies: The covariance matrix adaptation. In: International Conference on Evolutionary Computation, pp. 312–317 (1996)
2. Ledoit, O., Wolf, M.: Honey, i shrunk the sample covariance matrix. J. Portfolio Manag. **30**(4), 110–119 (2004)
3. Bishop, C.M.: Pattern Recognition and Machine Learning (Information Science and Statistics). Springer (2007)
4. Sharpe, W.F.: A simplified model for portfolio analysis. Manag. Sci. **9**(1), 277–293 (1963)
5. Brys, T., Drugan, M.M., Bosman, P.A.N., Cock, M.D., Nowé, A.: Solving satisfiability in fuzzy logics by mixing CMA-ES. In: Proceedings of the Genetic and Evolutionary Computation Conference, GECCO 2013, pp. 1125–1132 (2013)
6. Kruisselbrink, J.W., Reehuis, E., Deutz, A.H., Bäck, T., Emmerich, M.: Using the uncertainty handling CMA-ES for finding robust optima. In: Proceedings of the 13th Annual Genetic and Evolutionary Computation Conference, GECCO 2011, pp. 877–884. Dublin, Ireland, 12–16 July 2011
7. Chen, C., Chen, Y., Shen, T., Zao, J.K.: On the optimization of degree distributions in LT code with covariance matrix adaptation evolution strategy. In: Proceedings of the IEEE Congress on Evolutionary Computation, CEC 2010, pp. 1–8 (2010)
8. Igel, C., Suttorp, T., Hansen, N.: A computational efficient covariance matrix update and a (1+1)-CMA for evolution strategies. In: Proceedings of the Genetic and Evolutionary Computation Conference, GECCO 2006, pp. 453–460 (2006)
9. Suttorp, T., Hansen, N., Igel, C.: Efficient covariance matrix update for variable metric evolution strategies. Mach. Learn. **75**(2), 167–197 (2009)
10. Loshchilov, I.: A computationally efficient limited memory CMA-ES for large scale optimization. In: Proceedings of the Genetic and Evolutionary Computation Conference, GECCO 2014, pp. 397–404. Vancouver, BC, Canada, 12–16 July 2014
11. Caraffini, F., Iacca, G., Neri, F., Picinali, L., Mininno, E.: A CMA-ES super-fit scheme for the re-sampled inheritance search. In: Proceedings of the IEEE Congress on Evolutionary Computation, CEC 2013, pp. 1123–1130 (2013)
12. Rodrigues, S.M.F., Bauer, P., Bosman, P.A.N.: A novel population-based multi-objective CMA-ES and the impact of different constraint handling techniques. In: Proceedings of the Genetic and Evolutionary Computation Conference, GECCO 2014, pp. 991–998. Vancouver, BC, Canada, 12–16 July 2014
13. Santos, T., Takahashi, R.H.C., Moreira, G.J.P.: A CMA stochastic differential equation approach for many-objective optimization. In: Proceedings of the IEEE Congress on Evolutionary Computation, CEC 2012, pp. 1–6 (2012)
14. Voß, T., Hansen, N., Igel, C.: Improved step size adaptation for the MO-CMA-ES. In: Proceedings of the Genetic and Evolutionary Computation Conference, GECCO 2010, pp. 487–494 (2010)
15. Arnold, D.V., Hansen, N.: A (1+1)-CMA-ES for constrained optimisation. In: Proceedings of the Genetic and Evolutionary Computation Conference, GECCO 2012, pp. 297–304. Philadelphia, PA, USA, 7–11 July 2012
16. Loshchilov, I.: CMA-ES with restarts for solving CEC 2013 benchmark problems. In: Proceedings of the IEEE Congress on Evolutionary Computation, CEC 2013, pp. 369–376 (2013)
17. Beyer, H.G., Sendhoff, B.: Covariance matrix adaptation revisited—the cmsa evolution strategy. In: Proceedings of the 10th Conference on Parallel Problem Solving from Nature, PPSN X 2008, pp. 123–132 (2008)
18. Au, C., Leung, H.: Eigenspace sampling in the mirrored variant of $(1, \lambda)$-cma-es. In: Proceedings of the IEEE Congress on Evolutionary Computation, CEC 2012, pp. 1–8 (2012)
19. Krause, O., Glasmachers, T.: A CMA-ES with multiplicative covariance matrix updates. In: Proceedings of the Genetic and Evolutionary Computation Conference, GECCO 2015, pp. 281–288. Madrid, Spain, 11–15 July 2015

20. Ochoa, A.: Opportunities for expensive optimization with estimation of distribution algorithms. In: Tenne, Y., Goh, C.-K. (eds.) Computational Intelligence in Expensive Optimization Problems, pp. 193–218. Springer (2010)
21. Kanji, G.: 100 Statistical Tests. SAGE Publications, London (1993)

Part II
Machine Learning

Chapter 4
Machine Learning

4.1 Introduction

The overall amount of data is steadily growing. Examples are the human genome project, NASA earth observations, time series in smart power grids, and the enormous amount of social media data. Learning from data belongs to the most important and fascinating fields in computer science. The discipline is called machine learning or data mining. The reason for the fast development of machine learning is the enormous growth of data sets in all disciplines. For example in bioinformatics, large data sets of genome data have to be analyzed to detect illnesses and for the development of drugs. In economics, the analysis of large data sets of market data can improve the behavior of decision makers. Prediction and inference can help to improve planning strategies for efficient market behavior. The analysis of share markets and stock time series can be used to learn models that allow the prediction of future developments. There are thousands of further examples that require the development of efficient data mining and machine learning techniques. Machine learning tasks vary in various kinds of ways, e.g., the type of learning task, the number of patterns, and their size.

Learning means that new knowledge is generated from observations and that this knowledge is used to achieve defined objectives. Data itself is already knowledge. But for certain applications and for human understanding, large data sets cannot directly be applied in their raw form. Learning from data means that new condensed knowledge is extracted from the large amount of information. Various data learning tasks arise in machine learning. Prediction models can be learned with label information, data can be grouped without label information. Also visualization belongs to the important problem classes. For each task, numerous methods have been proposed and developed in the past decades. Some of them can be applied to a broad set of problems, while others are restricted to specific domains and applications. It is ongoing research to develop specialized methods for learning, pre- and post-processing for different domains. The corresponding machine learning process chain will be described in each chapter.

© Springer International Publishing Switzerland 2016
O. Kramer, *Machine Learning for Evolution Strategies*,
Studies in Big Data 20, DOI 10.1007/978-3-319-33383-0_4

This chapter will introduce the foundations of machine learning that are necessary for understanding the following chapters. It is structured as follows. First, the foundations of prediction and inference are introduced in Sect. 4.2. Section 4.3 gives an introduction to classification with a short overview of popular classifiers. Basic principles of model selection like cross-validation are introduced in Sect. 4.4. In high-dimensional data spaces, many problems suffer from the curse of dimensionality that is illustrated in Sect. 4.5. The bias-variance trade-off is presented in Sect. 4.6. Feature selection is an important pre-processing method and is described in Sect. 4.7. Conclusions are drawn in Sect. 4.8.

4.2 Prediction and Inference

Supervised learning means learning with labels. Labels are additional information we can use to train a model and to predict, if they are missing. Nominal patterns consist of variables likes names, which do not have a numerical interpretation. Interesting variants are ordinal variables, whose values are sorted, e.g., high, medium, and low. Numerical variables have continuous or integer values.

There are basically two types of supervised learning problems: prediction and inference. If some patterns $\mathbf{x}_i$ and corresponding labels y_i with $i = 1, \ldots, N$ are available, but labels are desired for new patterns that carry no labels, the problem is called a prediction problem. In a prediction problem, we seek for a model f that represents the estimate of a real but unknown function $\tilde{f}$, i.e., we seek for a model f with

$$y = f(\mathbf{x}). \tag{4.1}$$

Model f is our machine learning model that has to be learned from the observed data, y is the target value, which we denote as label. The machine learning model employs parameters that can be tuned during a training process. In the process of fitting f to $\tilde{f}$, two types of errors occur: the reducible error and the irreducible. The reducible error results from the degree of freedom of f when fitting to the observations coming from the true model. Adapting the free parameters of f reduces the reducible error during the training process. The irreducible error ϵ results from random errors in the true model $\tilde{f}$ that cannot be learned by the machine learning model. Error ϵ is the non-systematic part of the true model resulting in

$$y = f(\mathbf{x}) + \epsilon. \tag{4.2}$$

If we consider a fixed model f and the pattern-label pairs $(\mathbf{x}_1, y_1), \ldots, (\mathbf{x}_N, y_N)$, the expected test mean squared error (MSE) is

$$E(y - \tilde{f}(\mathbf{x}))^2 = \mathrm{Var}(\tilde{f}(\mathbf{x})) + \left(\mathrm{Bias}(\tilde{f}(\mathbf{x}))\right)^2 + \mathrm{Var}(\epsilon). \tag{4.3}$$

It corresponds to the test MSE we obtain, if we repeatedly estimate f on a large number of training sets, each tested on $\mathbf{x}$, see [1]. Var(ϵ) is the variance of the error term ϵ, while Bias($\tilde{f}(\mathbf{x})$) measures the bias of the true model $\tilde{f}$. For a discussion of the bias, also see Sect. 4.6. The first two terms are the reducible parts of the expected error, while Var(ϵ) is the irreducible part. Objective of the learning process is to minimize the reducible error of machine learning model f. However, the error of f can never be lower than the irreducible error, i.e., Var(ϵ) is an upper bound on the accuracy of f.

If we are interested in the way y is influenced by the patterns $\mathbf{x}_1, \ldots, \mathbf{x}_N$, the objective is called inference, in contrast to prediction. A relevant question in this context is, what the relationships between patterns and labels are. Is an attribute correlated to the label, and does, from a qualitative point of view, an increase of an attribute increase or decrease the label? Are all attributes correlated with the label or only a certain subset? This question is also known as feature selection problem.

4.3 Classification

Classification describes the problem of predicting discrete class labels for unlabeled patterns based on observations. Let $(\mathbf{x}_1, y_1), \ldots, (\mathbf{x}_N, y_N)$ be observations of d-dimensional continuous patterns, i.e., $\mathbf{x}_i \in \mathbb{R}^d$ with discrete labels $y_1, \ldots, y_N$. The objective in classification is to learn a functional model f that allows a reasonable prediction of unknown class labels y' for a new pattern $\mathbf{x}'$. Patterns without labels should be assigned to labels of patterns that are similar, e.g., that are close to the target pattern in data space, that come from the same distribution, or that lie on the same side of a separating decision function. But learning from observed patterns can be difficult. Training sets can be noisy, important features may be unknown, similarities between patterns may not be easy to define, and observations may not be sufficiently described by simple distributions. Further, learning functional models can be tedious task, as classes may not be linearly separable or may be difficult to separate with simple rules or mathematical equations.

Meanwhile, a large number of classification methods, called classifiers, has been proposed. The classifiers kNN and SVMs will be presented in more detail in Chaps. 6 and 7, where they have an important part to play in combination with the ES. An introduction of their working principles will follow in the corresponding chapters.

Closely related to nearest neighbor methods are kernel density methods. They compute labels according to the density of the patterns in the neighborhood of the requested patterns. Densities are measured with a kernel density function $K_\mathbf{w}$ using bandwidths $\mathbf{w}$ that define the size of the neighborhoods for the density computations. Kernel regression is also known as Nadaraya-Watson estimator and is defined as

$$f(\mathbf{x}') = \sum_{i=1}^{N} y_i \frac{K_\mathbf{w}(\mathbf{x}' - \mathbf{x}_i)}{\sum_{j=1}^{N} K_\mathbf{w}(\mathbf{x}' - \mathbf{x}_j)}. \tag{4.4}$$

The bandwidths control the smoothness of the regression function. Small values lead to an overfitted prediction function, while high values tend to generalize.

Neural networks belong to a further famous class of classification method. Similar to EAs, they belong to the bio-inspired machine learning methods. Neural networks use layers of neurons that represent non-linear functions. Their parameters, also known as weights, are adapted to reduce the training error, mostly by performing gradient descent in the space of weights. Recently, deep learning [2, 3] receives great attention. Deep neural networks use multiple layers with a complex structure. They turn out to be very successful in speech recognition, image recognition, and numerous other applications.

Decision trees are based on the idea to traverse a tree feature by feature until the leaf determines the label of the target pattern. Ensembles of decision trees are known as random forests. They combine the predictions of various methods to one prediction and exploit the fact that many specialized but overall weak estimators may be stronger than a single one. Bayes methods use the Bayes equation to compute the label given a training set of observations. With the assumption that all features come from different distributions, Naive Bayes computes the joint probability by multiplication of the single ones.

For more complex classification tasks, methods like SVMs [4, 5] are introduced. SVMs employ a linear decision boundary that separates different classes. To get a results that generalizes best, SVMs maximize the margin, i.e., the distance to the closest patterns to the decision boundary. To allow the classification of non-linearly separable patterns, slack variables are introduced that soften the constraint of separating patterns. Further, kernel functions map the data space to a feature space, where patterns are linearly separable. SVMs will shorty be presented and applied in Chap. 7.

Delgado et al. [6] compare numerous supervised learning methods experimentally, i.e., 179 classifiers from 17 families of methods on 121 data sets. The best results are reported for random forests and SVMs with RBF-kernel.

4.4 Model Selection

The problem in supervised learning is to find an adequate model f and to tune its parameters. A problem that may occur while training a method is overfitting. Overfitting means that a model fits well to the training data, but fails to achieve the same accuracy on an independent test data set. Cross-validation is a method to avoid overfitting. The idea of cross-validation is to split up the N observations $\{(\mathbf{x}_i, y_i)\}_{i=1}^{N}$ into training, validation, and test sets. The training set is used as basis for learning algorithms given a potential parameter set. The validation set is used to evaluate the model given the parameter set. The optimized model is finally evaluated on an independent test set that has not been used for training and validation of model f.

Minimization of the error on the validation set is basis of the training phase. For optimization of parameters, e.g., the parameters of an SVM, weak strategies like

Fig. 4.1 Illustration of
5-fold cross-validation. The
training set is divided into 5
folds. In 5 training settings,
the model is trained on 4
folds, and validated on the
one left out. The overall error
is aggregated to a
cross-validation score

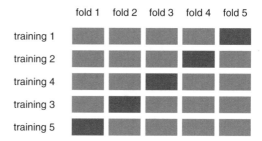

grid search are often sufficient. An advanced strategy to avoid overfitting is n-fold cross-validation that repeats the learning process n times with different training and validation sets. For this sake, the data set is split up into n disjoint sets, see Fig. 4.1 for $n = 5$ In each step, model f employs $n - 1$ sets for training and is evaluated on the remaining validation set. The error is aggregated to select the best parameters for model f on all n validation sets and is used to compute the cross-validation score. Advantage of this procedure is that all observations have been used for training the model and not only a subset of a small data set. In case of tiny data sets, the number of patterns might be too small to prevent that model f is not biased towards the training and validation set. In this case, the n-fold cross-validation variant $n = N$ called leave-one-out cross-validation (LOO-CV) is a recommendable strategy. In LOO-CV, one pattern is left out for prediction based on the remaining $N - 1$ training patterns. The whole procedure is repeated N times.

4.5 Curse of Dimensionality

Many machine learning methods have problems in high-dimensional data spaces. The reason is an effect also known as curse of dimensionality or Hughes effect. In high-dimensional data spaces, many patterns are required to cover the whole data space. If we distribute patterns randomly in the space along one axis, a quadratic number of patterns would be required to achieve the same coverage for a square with the same width. In general, an exponential number of patterns is required to cover a cube with increasing dimensions, and many machine learning methods require a good coverage of the data space with patterns.

Hastie et al. [4] present an interesting example for this effect. We assume points uniformly sampled in the unit hypercube. The volume of a d-dimensional unit hypercube is 1^d. Let r with $r < 1$ be the hypercube width of a smaller hypercube corresponding to the volume of the unit hypercube we want to capture. Then, r^d is its volume and at the same time the volume ratio v of the unit hypercube $v = r^d$. If we consider a smaller hypercube with width 0.8 in a 10-dimensional data space, we can interpret the situation as follows. We fill 80 % of each dimension of the

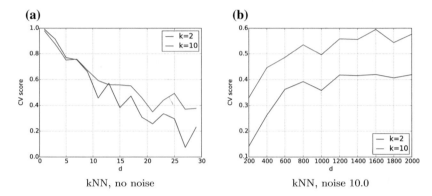

Fig. 4.2 Experimental illustration of the curse of dimensionality problem with kNN on a regression problem generated with the SCIKIT-LEARN method `make_regression`. **a** The kNN prediction score for 5-fold cross-validation significantly decreases with increasing dimension. **b** The model accuracy improves in the high-dimensional space, if the set of patterns is augmented

unit hypercube, but only cover $v = 0.8^{10} \approx 0.1$ of its volume, i.e., 90 % of the 10-dimensional hypercube is empty.

To demonstrate the curse of dimensionality problem in machine learning, we analyze kNN regression, which will be introduced in detail in Chap. 6, on a toy data set generated with the SCIKIT-LEARN method `make_regression`. Figure 4.2 shows the cross-validation regression score (the lower the better, see Chap. 5) for 5-fold cross-validation on a data set of size $N = 200$ with an increasing number of dimensions $d = 1, \dots, 30$. We can observe that the cross-validation score decreases with increasing number of dimensions. The reason is that the coverage of the data space with patterns drops significantly. The plot on the right shows how the score increases for $d = 50$, when the number of patterns is augmented from $N = 200$ to 2000. The choice $k = 10$ almost always achieves better results than $k = 2$. Averaging over the labels of more patterns obviously allows better generalization behaviors for this regression problem.

4.6 Bias-Variance Trade-Off

An inflexible model with few parameters may have problems to fit data well. But its sensibility to changes in the training set will be relatively moderate in comparison to a model that is very flexible with many parameters that allow an arbitrary adaptation to data space characteristics. Inflexible models have comparatively few shapes they can adopt, but are often easy to interpret. For example, linear models assume linear relationships between attributes, which are easy to describe with their coefficients, e.g.,

$$f(\mathbf{x}) = \beta_0 + \beta_1 x_1 + \dots + \beta_d x_d. \tag{4.5}$$

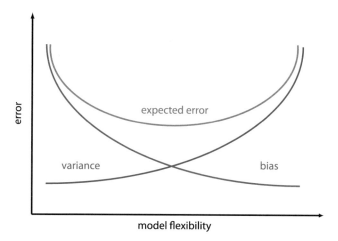

Fig. 4.3 Illustration of bias-variance trade-off

The coefficients β_i model the relationships and are easy to interpret. Optimizing the coefficients of a linear model is easier than fitting an arbitrary function with multiple coefficients. However, linear models will less likely suffer from overfitting as they are less depending on slight changes in the training set. They have low variance that is the amount by which the model changes, if using different training data sets. Such models have large errors when approximating a complex problem corresponding to a high bias. Bias is a measure for the inability of fitting the model to the training patterns. In contrast, flexible methods have high variance, i.e., they vary a lot when changing the training set, but have low bias, i.e., they better adapt to the observations. The bias and variance terms are shown in Eq. 4.3.

Figure 4.3 illustrates the bias-variance trade-off. On the x-axis the model complexity increases from left to right. While a method with low flexibility has a low variance, it usually suffers from high bias. The variance increases while the bias decreases with increasing model flexibility. The effect changes in the middle of the plot, where variance and bias cross. The expected error is minimal in the middle of the plot, where bias and variance reach a similar level. For practical problems and data sets, the bias-variance trade-off has to be considered when the decision for a particular method is made.

4.7 Feature Selection and Extraction

Feature selection is the process of choosing appropriate features from a set of available data. The choice of appropriate features for a certain machine learning task is an important problem. In many inference and prediction scenarios, not all features are relevant, but a subset. Other features may be redundant and strongly correlated. For example, attributes can be linear combinations of others. Only few dimensions might be sufficient to describe the distribution of data. An interesting question is, if a subset of features can achieve the same prediction accuracy as the full set. The classifier performance can also increase, when disturbing patterns are removed from the setting. Evolutionary methods are frequently applied for feature selection tasks. For example, in [7], we employ an evolutionary feature selection approach for data-driven wind power prediction. The approach makes use of a spatio-temporal regression approach and selects the best neighboring turbines with an EA and binary representation.

Feature extraction is related to feature selection. New features are generated from observed ones. In image analysis, the computation of color histograms and the detection of locations and frequencies of edges are examples for feature extraction. From high-dimensional patterns, meaningful new attributes are generated that capture important aspects to accomplish a task like characterization of human voices and phonemes in speech recognition. Features can also automatically be extracted with dimensionality reduction methods like PCA, which will be introduced in Chap. 8.

4.8 Conclusions

Supervised learning methods have grown to strong and successful tools for prediction and inference. With the steadily growing amount of data being collected in numerous disciplines, they have reached an outstanding importance. This chapter gives an introduction to important concepts in machine learning and presents foundations for the remainder of this book with an emphasis on supervised learning. Model selection, i.e., the choice and parameterization of appropriate methods, is an important topic. But to avoid overfitting, the methods have to be trained in a cross-validation setting. Cross-validation separates the pattern set into training and validation sets. Patterns from the training set serve as basis for the learning process, while the trained model is evaluated on the validation set. The independence of both sets forces the model to generalize on the training set and thus to adapt the parameters appropriately.

The curse of dimensionality problem complicates supervised learning tasks. With increasing dimensionality of the problem, the hardness significantly grows due to a sparser coverage of the data space with patterns. The collection of more patterns and the application of dimensionality reduction methods are techniques to solve the curse of dimensionality problem.

For a deeper introduction to machine learning, we refer to textbooks like Hastie et al. [4] and Bishop [8]. In general, an important aspect is to have strong methods at hand in form of implementations that can be used in a convenient kind of way. The PYTHON library for machine learning SCIKIT-LEARN offers an easy integration of techniques into own developments. It will be introduced in the following chapter.

References

1. James, G., Witten, D., Hastie, T., Tibshirani, R.: An Introduction to Statistical Learning. Springer, Heidelberg (2013)
2. Bengio, Y.: Learning deep architectures for AI. Found. Trends Mach. Learn. **2**(1), 1–127 (2009)
3. Deng, L., Yu, D.: Deep learning: methods and applications. Found. Trends Sig. Process. **7**(3–4), 197–387 (2014)
4. Hastie, T., Tibshirani, R., Friedman, J.: The Elements of Statistical Learning. Springer, Heidelberg (2009)
5. Vapnik, V.: The Nature of Statistical Learning Theory. Springer, New York (1995)
6. Delgado, M.F., Cernadas, E., Barro, S., Amorim, D.G.: Do we need hundreds of classifiers to solve real world classification problems? J. Mach. Learn. Res. **15**(1), 3133–3181 (2014)
7. Treiber, N.A., Kramer, O.: Evolutionary turbine selection for wind power predictions. In: Proceedings of the 37th Annual German Conference on AI, KI 2014: Advances in Artificial Intelligence, pp. 267–272. Stuttgart, Germany (2014)
8. Bishop, C.M.: Pattern Recognition and Machine Learning (Information Science and Statistics). Springer (2007)

Chapter 5
Scikit-Learn

5.1 Introduction

SCIKIT-LEARN is an open source machine learning library written in Python [1].
It allows the easy and fast integration of machine learning methods into PYTHON
code. The SCIKIT-LEARN library comprises a wide bandwidth of methods for classification, regression, covariance matrix estimation, dimensionality reduction, data
pre-processing, and benchmark problem generation. It can be accessed via the URL
http://scikit-learn.org. It is available for various operating systems and is easy to
install. The library is steadily improved and extended. SCIKIT-LEARN is widely used
in many commercial applications and is also part of many research projects and
publications. The SCIKIT-LEARN implementations are the basis of many methods
introduced in this book. To improve efficiency, some algorithms are implemented in
C and integrated via CYTHON, which is an optimizing static compiler for PYTHON
and allows easy extensions in C. The SVM variants are based on LIBSVM and
integrated into SCIKIT-LEARN with a CYTHON wrapper. LIBLINEAR is a library containing methods for linear classification. It is used to import linear SVMs and logistic
regression.

In Sect. 5.2, we will introduce methods for data management, i.e., loading and
generation of benchmark data sets. For demonstration purposes, we will employ
the solution candidates generated by an exemplary run of a (1+1)-ES on the constrained benchmark function, i.e., the Tangent problem, see Appendix A. The use of
supervised learning methods like nearest neighbors with model training and prediction is introduced in Sect. 5.3. Pre-processing methods like scaling, normalization,
imputation, and feature selection are presented in Sect. 5.4, while model evaluation
methods are sketched in Sect. 5.5. Section 5.6 introduces cross-validation methods.
Unsupervised learning is presented in Sect. 5.7, and conclusions are finally drawn in
Sect. 5.8.

© Springer International Publishing Switzerland 2016
O. Kramer, *Machine Learning for Evolution Strategies*,
Studies in Big Data 20, DOI 10.1007/978-3-319-33383-0_5

5.2   Data Management

An important part of SCIKIT-LEARN is data management, which allows loading typical machine learning data sets, e.g., from the UCI machine learning repository. An example is the Iris data set that contains 150 4-dimensional patterns with three classes that are types of Iris flowers. Data set modules are imported via `from sklearn import datasets`, while `iris = datasets.load_iris()` loads the Iris data set.

Another example is the California housing data set. The data set is represented by a dictionary-like data structure with the attributes `dataset.data`, which is a 20,640 times 8-shape `numpy` array. The array `dataset.target` contains the labels corresponding to the average house value in units of 100000. Numerous further data sets are available. The list is continuously extended.

For a convenient way of splitting the data set into training and test set, the method `train_test_split` from `sklearn.cross_validation` can be used. With parameter `test_size` the fraction of the test set size can be specified with a value between 0 and 1. For example, for a 80/20 split of training and test data, the setting `test_size = 0.2` must be used.

Further, SCIKIT-LEARN employs methods that allow generating data sets, so called sample generators. There exist regression and classification data set generators that allow the specification of interesting data set properties like sparsity, etc. Examples are `make_classification`, `make_circle`, `make_regression`, and `make_hastie`. For example, the latter generates a data set for binary classification. The output consists of 10 features with standard independent Gaussian labels, i.e.,

```
y[i] = 1 if np.sum(X[i] ** 2) > 9.34 else -1.
```

In Fig. 5.1, we will analyze a data set that has been generated during optimization of a (1+1)-ES on the Tangent problem with $N = 10$.

(a) 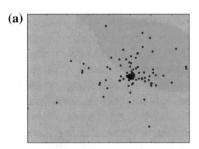   **(b)**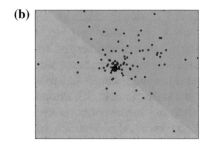

SVM on Tangent problem, RBF-kernel, standard settings,$C = 10^3$

SVM on Tangent problem tuned with cross-validation, resulting in RBF-kernel, $C = 10^7$, and $\gamma = 10^{-3}$

Fig. 5.1 SVM on the Tangent problem for $d = 2$ dimensions **a** with standard settings and **b** with optimized settings determined by cross-validation

5.3 Supervised Learning

In Chap. 4, the basis of machine learning is introduced. In Chap. 6, we will employ the first supervised learning algorithm, i.e., nearest neighbor regression. Here, we will show how to use the kNN implementation of SCIKIT-LEARN. Training, fitting, and prediction with the nearest neighbor method kNN works as follows.

1. The command `from sklearn import neighbors` imports the nearest neighbor package of SCIKIT-LEARN that contains the kNN classifier.
2. `clf = neighbors.KNeighborsClassifier(n_neighbors = 2, algorithm = 'ball_tree')` instantiates a kNN model with neighborhood size $k = 2$ employing a ball tree. Ball trees allows efficient neighborhood requests in $O(\log N)$ for low-dimensional data sets.
3. `clf.fit(X, y)` trains the classifier with patterns collected in `X` (which can be a list or a NUMPY array) and corresponding labels in `y`.
4. `clf.predict(X_)` finally applies to model to the patterns in `X_` yielding a corresponding list of labels.

Other classification and regression methods are used similarly, i.e., with the methods `fit` and `predict`. A famous method for supervised learning with continuous labels is linear regression that fits a linear model to the data. The command `regr = linear_model.LinearRegression()` creates a linear regression object. The method `regr.fit(X, y)` trains the linear regression model with training patterns `X` and labels `y`. Variants of linear models exist like ridge regression (`linear_model.Ridge`) and kernel ridge regression (`linear_model.KernelRidge`). A further prominent method is Lasso (`linear_model.Lasso`) that improves conditioning of the data by mitigating the curse of dimensionality. It selects only the informative features and sets the non-informative ones to zero. Lasso penalizes coefficients with L1 prior as regularizer. Decision trees are very successful methods and also part of the SCIKIT-LEARN library. With `tree.DecisionTreeClassifier()` a decision tree is available.

For support vector classification, the class `sklearn.svm.SVC` is implemented. It is based on the libraries LIBSVM and LIBLINEAR with the following interesting parameters. For example, parameter `C` is the regularization parameter that controls the penalty term of the SVM. With `kernel`, a kernel function can be specified. A default choice is a radial-basis function (RBF) kernel. In case of polynomial kernels, `degree` specifies its degree. Due to numerical reasons, the SVM training process may get stuck. Parameter `max_iter` allows the specification of a maximum number of training steps of the optimizer and forces the training to terminate. The standard setting (`max_iter = -1`) does not set a limit on the number of optimization steps.

Figure 5.1 shows the results when employing SVMs for classification of the feasibility of the constrained solution space when optimizing the Tangent problem with the (1+1)-ES as introduced in Chap. 2 for 200 generations.

5.4   Pre-processing Methods

Pre-processing of patterns is often required to allow successful learning. An essential part is scaling. For the successful application of many methods, data must look standard normally distributed with zero mean and unit variance. The pre-processing library is imported with the expression from sklearn import preprocessing. A numpy array X is scaled with

```
X_scaled = preprocessing.scale(X).
```

The result has zero mean and unit variance. Another scaling variant is the min-max scaler. Instantiated with

```
min_max_scaler = preprocessing.MinMaxScaler(),
```

the data can be scaled to values between 0 and 1 by

```
X_minmax = min_max_scaler.fit_transform(X)
```

for a training set X. Normalization is a variant that maps patterns so that they have unit norm. The assumption is useful for text classification and clustering and also supports pipelining. Feature transformation is an important issue. A useful technique is the binarization of features. Instantiated with

```
binarizer = preprocessing.Binarizer().fit(X),
```

the binarizer's method binarizer.transform(X) allows thresholding numerical features to get boolean features. This may be useful for many domains, e.g., text classification.

Besides binary or numerical features, we can have categorical features. SCIKIT-LEARN features the one-hot encoder available via preprocessing.OneHot Encoder(). It encodes categorical integer features employing a one-of-K scheme, i.e., outputs a sparse matrix, where each column corresponds to a value of a feature. An input matrix of integers is required within a range based on the number of features.

A further class of pre-processing methods cares for imputation. In practical machine learning problems, patterns are often incomplete. Methods can only cope with complete patterns. If the removal of incomplete patterns is not possible, e.g., due small training sets, missing elements have to be filled with appropriate values. This is also possible with regression methods. SCIKIT-LEARN features the imputer that completes missing data (preprocessing.Imputer). A further feature transformation method is the method for polynomial features. It adds complexity to the model by considering nonlinear features similar to kernel functions.

Feature selection methods can also be seen as a kind of pre-processing step. The task is to select a subset of features to improve the classifier accuracy. Feature selection will also improve the speed of most classifiers. The method `feature_selection.SelectKBest` selects the K best features w.r.t. measures like `f_classif` for the ANOVA F-value between labels for classification tasks and `chi2` for Chi-squared statistics of non-negative features. Further measures are available. An interesting methods for feature selection is recursive feature elimination (`feature_selection.RFE`). It step by step removes features that are assigned by a classifier with a minimal weight. This process is continued until the desired number is reached. Besides feature selection, the concatenation of features is a useful method and easily possible via `feature_selection.FeatureUnion(X1, X2)`, where `X1` and `X2` are patterns.

SCIKIT-LEARN allows pipelining, i.e., the combination of multiple methods successively. An exemplary pipeline combines normalization and classification. With `pipeline.Pipeline` the pipeline object is loaded. For example, for instantiated components `normalization` and `KNNClassifier` the command

```
pipe = Pipeline(steps = [('norm',normalization),('KNN',
KNNClassifier)])
```

implements a pipeline with two steps. To be applicable to be a pipeline member in the middle of a pipeline, the method must implement the `transform` command. Pipelines are efficient expression for such successive pattern-wise processing steps.

5.5 Model Evaluation

Model evaluation has an important part to play in machine learning. The quality measure depends on the problem class. In classification, the precision score is a reasonable measure. It is the ratio of true positives (correct classification for pattern of class *positive*) and all positives (i.e., the sum of true positive and false positives). Intuitively, precision is the ability of a classifier not to label a negative pattern as positive. The precision score is available via `metrics.precision_score` with variants that globally count the total number of true positives, false negatives, and false positives (`average = 'micro'`). A further variant computes the matrix for each label with unweighted mean (`average = 'macro'`), or weighed by the support mean (`average = 'weighted'`) for taking into account imbalanced data.

Another important measure is recall (`metrics.recall_score`) that is defined as the ratio between the true positives and the sum of the true positives and the false negatives. Intuitively, recall is a measure for the ability of the classifier to find all the positive samples. A combination of the precision score and recall is

Table 5.1 Classification report for constraints

Domain	Precision	Recall	F1	Support
0	0.92	1.00	0.96	68
1	1.00	0.14	0.25	7
Avg/total	0.93	0.92	0.89	75

the F1 score (`metrics.f1_score`) that is a weighted average of both measures. All three measures can be generated at once with the classification report (`metrics.classification_report`). An example is presented in Table 5.1, which shows the corresponding measures for a data set generated by a (1+1)-ES, which runs on the Tangent problem. The (1+1)-ES uses Gaussian mutation and Rechenberg's step size control. The optimization runs results in a 10-dimensional data set comprising 200 patterns. The `classification_report` will also be used in Chap. 7. For regression, there are also metrics available like the R2 score (`metrics.r2_score`), whose best value is 1.0 and that can also return negative values for bad regression methods.

5.6   Model Selection

In Chap. 4, the importance of model selection has been emphasized. In SCIKIT-LEARN, numerous methods are implemented for model selection. Cross-validation is a method to avoid overfitting and underfitting. It is combined with grid search or other optimization methods for tuning the models' parameters. For many supervised methods, grid search is sufficient, e.g., for the hyper-parameters of SVMs. Among the cross-validation methods implemented in SCIKIT-LEARN is n-fold cross-validation (`cross_validation.kFold`), which splits the data set into n folds, trains on $n-1$ folds and tests on the left-out fold. As a variant, stratified n-fold cross-validation (`cross_validation.StratifiedKFold`) preserves the class ratios within each fold. If few training patterns are available, leave-one-out cross-validation should be used, available as `cross_validation.LeaveOneOut`. All cross-validation variants employ the parameter `shuffle` that is a boolean variable stating, if the data set should be shuffled (`shuffle = True`) or not `shuffle = False` before splitting the data set into folds. A useful object in SCIKIT-LEARN is the grid search object `GridSearchCV` that implements an n-fold cross-validation and gets a classifier, a dictionary of test parameters, the number of folds and a quality measure as arguments. It performs the cross-validation with the employed method and returns a classifier that achieves an optimal score as well as scores for all parameter combinations.

5.7 Unsupervised Learning

Unsupervised learning is learning without label information. Various methods for unsupervised learning are part of SCIKIT-LEARN. One important unsupervised problem is clustering, which is the task of grouping data sets w.r.t. their intrinsic properties. It has numerous applications. A famous clustering method is k-means, which is also implemented in SCIKIT-LEARN. Given the number k of clusters, k-means iteratively places the k clusters in data space by successively assigning all patterns to the closest cluster center and computing the mean of these clusters. With `cluster.KMeans`, k-means can be applied stating the desired number of cluster centers k. Fitting k-means to a set of patterns and getting the cluster labels is possible with `y = KMeans(n_clusters=2).fit_predict(X)`. To estimate an appropriate number k, i.e., `n_clusters`, the percentage of inner cluster variance can be plotted for different k. It is recommendable to choose k from the area of the largest change of slope of this variance curve. Various other methods for clustering are available in SCIKIT-LEARN like DBSCAN. DBSCAN, which can be accessed with `cluster.DBSCAN`, determines core samples of high density defined via the number `min_samples` of neighboring patterns within a radius `eps`. It expands clusters from the core samples to find further corse samples. Corner samples have less patterns than `min_samples` in their radius, but are located in the radius of core samples. The method is used with command `DBSCAN(eps = 0.3, min_samples = 10).fit(X)`. DBSCAN will be introduced in Chap. 10, where it is used for clustering-based niching.

To give a further demonstration of PYTHON code, the following steps show the generation of a list of clusters, which contain their assigned patterns.

```
label_list = k_means.fit_predict(X)
labels = list(set(labellist))
clusters = [[] for i in range(len(labels))]
for i in xrange(len(X)):
>>> clusters[labellist[i]].append(np.array(X[i]))
```

First, the list of labels are accessed from the trained k-means method. With the `set`-method, this list is cast to a set that contains each label only once. The third step generates a list of empty cluster lists, which is filled with the corresponding patterns in the `for`-loop.

The second important class of unsupervised learning is dimensionality reduction. PCA is a prominent example and very appropriate to linear data. In SCIKIT-LEARN, PCA is implemented with singular value decomposition from the linear algebra package `scipy.linalg`. It keeps only the most significant singular vectors for the projection of the data to the low-dimensional space. Available in `decomposition.PCA` with a specification of the target dimensionality `PCA(n_components = 2)`, again the method `fit(X)` fits the PCA to the data, while `transform(X)` delivers the low-dimensional points. A combination of both

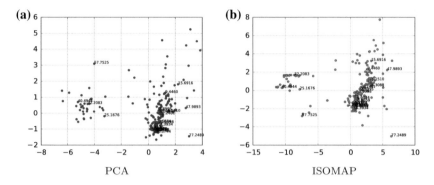

Fig. 5.2 a PCA embedding of 10-dimensional patterns of a (1+1)-ES for 200 generations.
b ISOMAP embedding of the corresponding data set with $k = 10$

methods is `fit_transform(X)`, which directly delivers the low-dimensional
pendants of the high-dimensional patterns. With `sklearn.manifold`, various
manifold learning methods for non-linear data are available. For example, ISOMAP
can be accessed via `manifold.Isomap` with parameters `n_neighbors` specify-
ing the number of neighbors employed for ISOMAP's neighborhood graph. The target dimensionality can again be set with `n_components`. With `fit_transform`
`(X)`, ISOMAP is fit to the data `X`, and the low-dimensional representations are
computed.

Figure 5.2 shows an application of PCA and ISOMAP optimizing the Tangent
problem. The corresponding data set is generated as follows. The (1+1)-ES runs for
200 generations resulting in 200 patterns. For dimensionality reduction, no labels
are employed, i.e., the fitness values are ignored. PCA and ISOMAP map into a
two-dimensional space for visualization. ISOMAP uses the neighborhood size $k =$
10. Both figures show that the optimization process converges towards a region in
solution space.

5.8 Conclusions

PYTHON is a modern programming language that allows fast prototyping of methods. It is a universal programming language, which is in most cases interpreted and
not compiled. PYTHON supports various programming paradigm like functional and
objective programming. PYTHON code is usually easy to read and shorter than code
in other high-level programming languages. This is achieved with a reduced number
of code words and a reduced syntax that allows short expressions. PYTHON supports dynamic typing of variables. Numerous libraries make PYTHON an attractive
programming language for a broad field of applications and domains.

SCIKIT-LEARN is a strong machine learning library written in PYTHON that developed fast in the past years. It allows an easy use of known machine learning concepts employing various important algorithms for pre-processing, feature selection, pipelining, model evaluation, and many other tasks. Numerous methods for supervised and unsupervised learning are implemented. This list of supported methods comprises nearest neighbor methods, SVMs, support vector regression (SVR), decision trees, random forests. Also unsupervised methods are available like k-means and DBSCAN for clustering and PCA, ISOMAP, and LEE for dimensionality reduction.

This chapter gives a short introduction and comprehensive overview over most of the concepts that are supported. SCIKIT-LEARN advertises that it is used by numerous companies and research institutes, e.g., Infria, Evernote, Spotify, and DataRobot. Further, SCIKIT-LEARN offers a very comprehensive documentation of the whole library and an excellent support by the developer and research community. In the course of this book, various techniques that are part of SCIKIT-LEARN are combined with the (1+1)-ES. Each chapter will give a short introduction to the relevant SCIKIT-LEARN method.

Reference

1. Pedregosa, F., Varoquaux, G., Gramfort, A., Michel, V., Thirion, B., Grisel, O., Blondel, M., Prettenhofer, P., Weiss, R., Dubourg, V., Vanderplas, J., Passos, A., Cournapeau, D., Brucher, M., Perrot, M., Duchesnay, E.: Scikit-learn: machine learning in Python. J. Mach. Learn. Res. **12**, 2825–2830 (2011)

Part III
Supervised Learning

Chapter 6
Fitness Meta-Modeling

6.1 Introduction

In expensive optimization problems, the reduction of the number of fitness function calls has an important part to play. Evolutionary operators produce candidate solutions $\mathbf{x}_i$ in the solution space that are evaluated on fitness function f. If a regression method $\hat{f}$ is trained with the pattern-label pairs $(\mathbf{x}_i, f(\mathbf{x}_i))$, $i = 1, \ldots, N$, the model $\hat{f}$ can be used to interpolate and eventually extrapolate the fitness of a novel solution $\mathbf{x}'$ that has been generated by the evolutionary operators. Model $\hat{f}$ is also known as meta-model or surrogate in this context. Many different regression methods can be employed for this purpose. An important question is how to manage the meta-model. Past fitness function evaluations are stored in a training set. The questions come up, which patterns to use for training the meta-model, how often to tune the meta-model, and when to use the fitness function or the meta-model as surrogate. Some methods directly use the fitness evaluations of the meta-model for the evolutionary process. Others use the meta-model for pre-screening, i.e., each solution is first evaluated on the meta-model and the most successful ones are evaluated on the real fitness function before being selected as parents for the following generation.

In this chapter, we analyze nearest neighbor regression when optimizing with a (1+1)-ES and Rechenberg's step size control technique. Nearest neighbor regression makes use of the idea that the label of an unknown pattern $\mathbf{x}'$ should get the label of the k closest patterns in the training set. This method is also known as instance-based and non-parametric approach. It does not induce a functional model like linear regression. The objective of this chapter is to show that a comparatively simple hybridization can result in a very effective optimization strategy. The chapter is structured as follows. Nearest neighbor regression is introduced in Sect. 6.2. The integration of the kNN meta-model is presented in Sect. 6.3. Related work is discussed in Sect. 6.4. The approach is experimentally analyzed in Sect. 6.5. Conclusions are drawn in Sect. 6.6.

© Springer International Publishing Switzerland 2016
O. Kramer, *Machine Learning for Evolution Strategies*,
Studies in Big Data 20, DOI 10.1007/978-3-319-33383-0_6

6.2 Nearest Neighbors

Nearest neighbor regression, also known as kNN regression, is based on the idea that the closest patterns to a target pattern $\mathbf{x}'$, for which we seek the label, deliver useful information for completing it. Based on this idea, kNN assigns the class label of the majority of the k-nearest patterns in data space. For this sake, we have to be able to define a similarity measure in data space. In $\mathbb{R}^d$, it is reasonable to employ the Minkowski metric (p-norm)

$$\|\mathbf{x}' - \mathbf{x}_j\|_p = \left(\sum_{i=1}^{d} |(x_i)' - (x_i)_j|^p \right)^{1/p} \tag{6.1}$$

with parameter $p \in \mathbb{N}$. The distance measure corresponds to the Euclidean distance for $p = 2$ and the Manhattan distance for $p = 1$. In other data spaces, adequate distance functions have to be chosen, e.g., the Hamming distance in $\mathbb{B}^d$. For regression tasks, kNN can also be applied. As continuous variant, the task is to learn a function $\hat{f} : \mathbb{R}^d \to \mathbb{R}$ known as regression function. For an unknown pattern $\mathbf{x}'$, kNN regression computes the mean of the function values of its k-nearest neighbors

$$\hat{f}(\mathbf{x}') = \frac{1}{K} \sum_{i \in \mathcal{N}_K(\mathbf{x}')} y_i \tag{6.2}$$

with set $\mathcal{N}_K(\mathbf{x}')$ containing the indices of the k-nearest neighbors of pattern $\mathbf{x}'$ in the training data set $\{(\mathbf{x}_i, y_i)\}_{i=1}^{N}$. Normalization of patterns is usually applied before the machine learning process, e.g., because different variables can come in different units.

The choice of k defines the locality of kNN. For $k = 1$, little neighborhoods arise in regions, where patterns from different classes are scattered. For larger neighborhood sizes, e.g. $k = 20$, patterns with labels in the minority are ignored. Neighborhood size k is usually chosen with the help of cross-validation. For the choice of k, grid-search or testing few typical choices like $[1, 2, 5, 10, 20, 50]$ may be sufficient. This restriction reduces the effort for tuning the model significantly. Nearest neighbor methods are part of the SCIKIT-LEARN package.

- The command `from sklearn import neighbors` imports the SCIKIT-LEARN implementation of kNN.
- `clf = neighbors.KNeighborsRegressor(n_neighbors=k)` calls kNN with k neighbors using uniform weights bei default. Optionally, an own distance function can be defined.
- `clf.fit(X,y)` trains kNN with patterns X and corresponding labels y.
- `y_pred = clf.predict(X_test)` predicts the class labels of test data set X_test.

The method kNN demonstrates its success in numerous applications, from classification of galaxies in digital sky surveys to the problem of handwritten digits and EKG data [1]. As kNN uses the training points that are nearest to the target patterns, kNN employs high variance and low bias, see the discussion in Chap. 4. For example, if the target patterns lie at the location of a training pattern, the bias is zero. Cover and Hart [2] show that the error rate of kNN with $k = 1$ is asymptotically bound by twice the Bayes error rate. The proof of this result is also sketched in Hastie et al. [1].

6.3   Algorithm

In this section, we introduce the meta-model-based ES. The main ingredients of meta-model approaches are the training set, which stores past fitness function evaluations, the meta-model maintenance mechanism, e.g., for parameter tuning and regularization, and the meta-model integration mechanism that defines how it is applied to save fitness function evaluations. Figure 6.1 illustrates the meta-model principle. Solutions are evaluated on the real fitness function f resulting in the blue squares. The meta-model is trained with these examples resulting in the red curve. This is basis of the meta-model evaluation of solutions, represented as red little squares.

One alternative to use the meta-model is to test each candidate solution with a certain probability on the meta-model and to use the predicted value instead of the real fitness function evaluation in the course of the evolutionary optimization process. We employ a different meta-model management that is tailored to the (1+1)-ES.

Algorithm 4 shows the pseudocode of the (1+1)-ES with meta-model (MM-ES) and Rechenberg's adaptive step size control [3]. If the solution $\mathbf{x}'$ has been evaluated on f, both will be combined to a pattern and as pattern-label pair $(\mathbf{x}', f(\mathbf{x}'))$ included to the meta-model training set. The last N solutions and their fitness function evaluations build the training set $\{(\mathbf{x}_i, f(\mathbf{x}_i))\}_{i=1}^{N}$. After each training set update, model $\hat{f}$ can be re-trained. For example in case of kNN, a new neighborhood size k may be chosen with cross-validation.

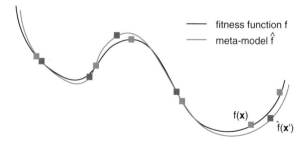

Fig. 6.1 Illustration of fitness meta-model. Solutions (*blue squares*) like $\mathbf{x}$ are evaluated on f (*black curve*). The meta-model $\hat{f}$ (*red curve*) is trained with the evaluated solutions. It estimates the fitness $\hat{f}(\mathbf{x}')$ of new candidate solutions (*red squares*) like $\mathbf{x}'$

Algorithm 4 MM-ES

1: initialize $\mathbf{x}$
2: **repeat**
3:    adapt σ with Rechenberg
4:    $\mathbf{z} \sim \sigma \cdot \mathcal{N}(\mathbf{0}, \mathbf{I})$
5:    $\mathbf{x}' = \mathbf{x} + \mathbf{z}$
6:    **if** $\hat{f}(\mathbf{x}') \leq f(\mathbf{x}_{-t})$ **then**
7:       evaluate $\mathbf{x}' \rightarrow f(\mathbf{x}')$
8:       last N solutions $\rightarrow \{(\mathbf{x}_i, f(\mathbf{x}_i))\}_{i=1}^{N}$
9:       train $\hat{f}$
10:       **if** $f(\mathbf{x}') \leq f(\mathbf{x})$ **then**
11:          replace $\mathbf{x}$ with $\mathbf{x}'$
12:       **end if**
13:    **end if**
14: **until** termination condition

The meta-model integration we employ is based on the idea that solutions are only evaluated, if they are promising. Let $\mathbf{x}$ be the solution of the last generation and let $\mathbf{x}'$ be the novel solution generated with the mutation operator. If the fitness prediction $\hat{f}(\mathbf{x}')$ of the meta-model indicates that $\mathbf{x}'$ employs a better fitness than the tth last solution $\mathbf{x}_{-t}$ that has been generated in the past evolutionary optimization progress, the solution is evaluated on the real fitness function f. The tth last solution defines a fitness threshold that assumes $\hat{f}(\mathbf{x}')$ may underestimate the fitness of $\mathbf{x}'$. The evaluations of candidate solutions that are worse than the threshold are saved and potentially lead to a decrease of the number of fitness function evaluations. Tuning of the model, e.g., the neighborhood size k of kNN with cross-validation, may be reasonable in certain optimization settings.

Last, the question for the proper regression model has to be answered. In our blackbox optimization scenario, we assume that we do not know anything about the curvature of the fitness function. For example, it is reasonable to employ a polynomial model in case of spherical fitness function conditions. But in general, we cannot assume to have such information.

6.4   Related Work

Meta-models, also known as surrogates, are prominent approaches to reduce the number of fitness function evaluations in evolutionary computation. Most work in meta-modeling concentrates on fitness function surrogates, few also on reducing constraint functions evaluations. In this line of research, early work concentrates on neural networks [4] and on Kriging [5]. Kriging belongs to the class of Gaussian process regression models and is an interpolation method. It uses covariance information and is based on piecewise-polynomial splines. Neural networks and Kriging meta-models are compared in [6]. An example for the recent employment of Kriging models is the differential evolution approach by Elsayed et al. [7].

Various kinds of mechanisms allow the savings of fitness function evaluations. Cruz-Vega et al. [8] employ granular computing to cluster points and adapt the parameters with a neuro-fuzzy network. Verbeeck et al. [9] propose a tree-based meta-model and concentrate on multi-objective optimization. Martínez and Coello [10] also focus on multi-objective optimization while employing a support vector regression meta-model. Loshchilov et al. [11] combine a one-class SVM with a regression approach as meta-model in multi-objective optimization. Ensembles of support vector methods are also used for in the approach by Rosales-Pérez [12] in multi-objective optimization settings. Ensembles combine multiple classifiers to reduce the fitness prediction error.

Kruisselbrink et al. [13] apply the Kriging model in CMA-ES-based optimization. The approach puts an emphasis on the generation of archive points for improving the meta-model. Local meta-models for the CMA-ES are learned in the approach by Bouzarkouna et al. [14], who train a full quadratic local model for each sub-function in each generation. Also Liao et al. [15] propose a locally weighted meta-model, which only evaluates the most promising candidate solutions. The local approach is similar to the nearest neighbor method we use in the experimental part, as kNN is a local method.

There is a line of research that concentrates on surrogate-assisted optimization for the CMA-ES. For example, the approach by Loshchilov et al. [16] adjusts the life length of the current surrogate model before learning a new surrogate as well as its hyper-parameters. A variant with larger population sizes [17] leads to a more intensive exploitation of the meta-model.

Preuss et al. [18] propose to use a computationally cheap meta-model of the fitness function and tune the parameters of the evolutionary optimization approach on this surrogate. Kramer et al. [19] combine two nearest neighbor meta-models, one for the fitness function, and one for the constraint function with an adaptive penalty function in a constrained continuous optimization scenario.

Most of the work sketched here positively reported savings in fitness function evaluations, although machine learning models and meta-model managing strategies vary significantly.

6.5 Experimental Analysis

In this section, we experimentally analyze the meta-model-based ES. Besides the convergence behavior, we analyze the influence of the neighborhood size k and the training set size N. Table 6.1 shows the experimental results of 25 runs of the (1+1)-ES and the MM-ES on the Sphere function and on Rosenbrock for kNN as meta-model with $k = 1$. Both algorithms get a budget of 5000 function evaluations. The results show the mean values and the corresponding standard deviations. On the Sphere function, we employ the setting $t = 10$. We can observe that the ES with

Table 6.1 Experimental comparison of (1+1)-ES and the MM-ES on the Sphere function and on Rosenbrock

Problem		(1+1)-ES		MM-ES		Wilx.
	d	Mean	Dev	Mean	Dev	p-value
Sphere	2	2.067e-173	0.0	2.003e-287	0.0	0.0076
	10	1.039e-53	1.800e-53	1.511e-62	2.618e-62	0.0076
Rosenbrock	2	0.260	0.447	8.091e-06	7.809e-06	0.0076
	10	0.519	0.301	2.143	2.783	0.313

meta-model leads to a significant reduction of fitness function evaluations. For $d = 2$, the standard deviation even falls below a value that is measurable due to a limited machine accuracy. The results are statistically significant, which is confirmed by the Wilcoxon test. On Rosenbrock, we set $t = 50$. The MM-ES significantly outperforms the (1+1)-ES for $d = 2$, but only in few runs on Rosenbrock leading to no statistical superiority for $d = 10$.

Figure 6.2 compares the evolutionary runs of the (1+1)-ES and the MM-ES on the Sphere function for (a) $d = 2$ and (b) $d = 10$. As of the very beginning of the optimization process, even the worst evolutionary runs with meta-model are better than the best evolutionary runs of the (1+1)-ES without meta-model. This effect is even more significant for $d = 2$.

Now, we analyze the training set size N as it is has a significant influence on the quality of regression model $\hat{f}$. In Fig. 6.3, we compare the two training set sizes $N = 20$ and $N = 500$ on the Sphere function. Again, we use the same settings, i.e., 5000 fitness function evaluations in each run for $d = 2$ and $d = 10$. The runs show that a too small training set lets the search become less stable in both cases. Some runs may get stuck because of inappropriate fitness function estimates. Further analyses show that a training set size of $N = 500$ is an appropriate choice.

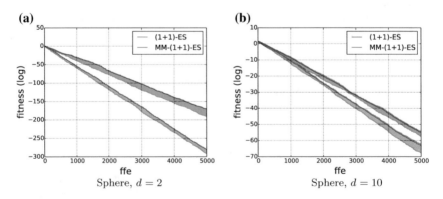

Fig. 6.2 Comparison of (1+1)-ES and MM-ES on the Sphere function with **a** $d = 2$ and **b** $d = 10$

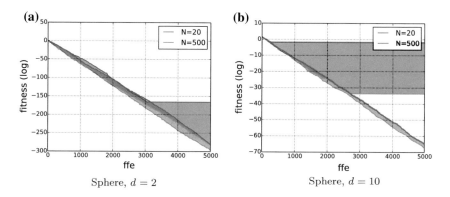

Fig. 6.3 Comparison of meta-model sizes $N = 20$ and $N = 500$ on the Sphere function with **a** $d = 2$ and **b** $d = 10$

Our analysis of the neighborhood size k have shown that the choice $k = 1$ yields the best results in all cases. Larger choices slow down the optimization or let the optimization process stagnate, similar to the stagnation we observe for small training set sizes. Hence, we understand the nearest neighbor regression meta-model with $k = 1$ as local meta-model, which also belongs to the most successful in literature, see Sect. 7.4.

Figure 6.4 shows the experimental results of the MM-ES on the Cigar function and on Rosenbrock for $d = 10$ dimensions and 5000 fitness function evaluations. It confirms that the MM-ES reduces the number of fitness function evaluations in comparison to the standard (1+1)-ES.

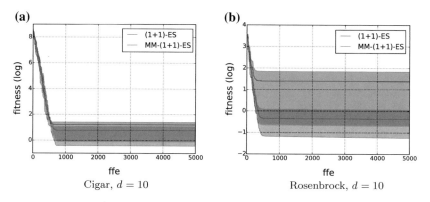

Fig. 6.4 Analysis of MM-ES on **a** Cigar and **b** on Rosenbrock for $d = 10$

6.6 Conclusions

Meta-modeling in evolutionary optimization is a frequent and well-known technique that allows saving expensive fitness function evaluations. In this chapter, we apply nearest neighbor regression in a (1+1)-ES with Gaussian mutation and Rechenberg's adaptive step size control. Fitness function evaluations are saved, if the predicted fitness is worse than the tth best element in the training set. The comparison with the tth best element of the past search and not with the best element is a pessimistic perspective. The search might get stuck or at least decelerate, if the model's quality is overestimated. Further, it turns out that the small neighborhood size $k = 1$ is the best choice. In our experiments, we observe significant performance wins on the Sphere function. The optimization runs of the (1+1)-ES with meta-model outperform the native ES. The Wilcoxon test confirms the statistical significance of all results.

Meta-models as fitness function surrogates can lead to various difficulties. The training set should be large enough to offer diversity that can be exploited for fitness predictions during the evolutionary optimization process. However, one has to keep in mind that inexact fitness predictions based on bad surrogates may disturb the evolutionary process and result in deteriorations instead of improvements. Various other ways to employ a regression model as meta-model are possible instead of pre-selection. For example, an interesting approach is the use of the meta-model as fitness function with a certain probability, but without further checks on the real fitness function.

References

1. Hastie, T., Tibshirani, R., Friedman, J.: The Elements of Statistical Learning. Springer, New York (2009)
2. Cover, T., Hart, P.: Nearest neighbor pattern classification **13**, 21–27 (1967)
3. Rechenberg, I.: Evolutionsstrategie - Optimierung technischer Systeme nach Prinzipien der biologischen Evolution. Frommann-Holzboog, Stuttgart (1973)
4. Jin, Y., Olhofer, M., Sendhoff, B.: On evolutionary optimization with approximate fitness functions. In: Proceedings of the Genetic and Evolutionary Computation Conference, GECCO 2000, pp. 786–793 (2000)
5. Armstrong, M.: Basic Linear Geostatistics. Springer (1998)
6. Willmes, L., Bäck, T., Jin, Y., Sendhoff, B.: Comparing neural networks and kriging for fitness approximation in evolutionary optimization. In: Proceedings of the IEEE Congress on Evolutionary Computation, CEC 2003, pp. 663–670 (2003)
7. Elsayed, S.M., Ray, T., Sarker, R.A.: A surrogate-assisted differential evolution algorithm with dynamic parameters selection for solving expensive optimization problems. In: Proceedings of the IEEE Congress on Evolutionary Computation, CEC 2014, pp. 1062–1068 (2014)
8. Cruz-Vega, I., Garcia-Limon, M., Escalante, H.J.: Adaptive-surrogate based on a neuro-fuzzy network and granular computing. In: Proceedings of the Genetic and Evolutionary Computation Conference, GECCO 2014, pp. 761–768 (2014)
9. Verbeeck, D., Maes, F., Grave, K.D., Blockeel, H.: Multi-objective optimization with surrogate trees. In: Proceedings of the Genetic and Evolutionary Computation Conference, GECCO 2013, pp. 679–686 (2013)

10. Martínez, S.Z., Coello, C.A.C.: A multi-objective meta-model assisted memetic algorithm with non gradient-based local search. In: Proceedings of the Genetic and Evolutionary Computation Conference, GECCO 2010, pp. 537–538 (2010)
11. Loshchilov, I., Schoenauer, M., Sebag, M.: A mono surrogate for multiobjective optimization. In: Proceedings of the Genetic and Evolutionary Computation Conference, GECCO 2010, pp. 471–478 (2010)
12. Rosales-Pérez, A., Coello, C.A.C., Gonzalez, J.A., García, C.A.R., Escalante, H.J.: A hybrid surrogate-based approach for evolutionary multi-objective optimization. In: Proceedings of the IEEE Congress on Evolutionary Computation, CEC 2013, pp. 2548–2555 (2013)
13. Kruisselbrink, J.W., Emmerich, M.T.M., Deutz, A.H., Bäck, T.: A robust optimization approach using kriging metamodels for robustness approximation in the CMA-ES. In: Proceedings of the IEEE Congress on Evolutionary Computation, CEC 2010, pp. 1–8 (2010)
14. Bouzarkouna, Z., Auger, A., Ding, D.Y.: Local-meta-model CMA-ES for partially separable functions. In: Proceedings of the Genetic and Evolutionary Computation Conference, GECCO 2011, pp. 869–876 (2011)
15. Liao, Q., Zhou, A., Zhang, G.: A locally weighted metamodel for pre-selection in evolutionary optimization. In: Proceedings of the IEEE Congress on Evolutionary Computation, CEC 2014, pp. 2483–2490 (2014)
16. Loshchilov, I., Schoenauer, M., Sebag, M.: Self-adaptive surrogate-assisted covariance matrix adaptation evolution strategy. In: Proceedings of the Genetic and Evolutionary Computation Conference, GECCO 2012, pp. 321–328 (2012)
17. Loshchilov, I., Schoenauer, M., Sebag, M.: Intensive surrogate model exploitation in self-adaptive surrogate-assisted cma-es (saacm-es). In: Proceedings of the Genetic and Evolutionary Computation Conference, GECCO 2013, pp. 439–446 (2013)
18. Preuss, M., Rudolph, G., Wessing, S.: Tuning optimization algorithms for real-world problems by means of surrogate modeling. In: Proceedings of the Genetic and Evolutionary Computation Conference, GECCO 2010, pp. 401–408 (2010)
19. Kramer, O., Schlachter, U., Spreckels, V.: An adaptive penalty function with meta-modeling for constrained problems. In: Proceedings of the IEEE Congress on Evolutionary Computation, CEC 2013, pp. 1350–1354 (2013)

Chapter 7
Constraint Meta-Modeling

7.1 Introduction

In practical optimization problems, constraints play an important role. Constraints decrease the allowed solution space to a feasible subset of the original one. Similar to the fitness function, we assume that fitness and constraint functions are blackboxes, i.e., nothing is known about the functions but evaluations via function calls with candidate solutions $\mathbf{x}$. The constraint function is usually denoted as $g(\mathbf{x})$ in literature and also in this chapter.

We define the constrained optimization problem as the problem to minimize $f(\mathbf{x})$ subject to the inequality constraints $g_i(\mathbf{x})$ with $i = 1, \ldots, n_c$. Inequality constraints can use the operators $\geq$ or $\leq$. If we define the constraints as $g_i(\mathbf{x}) \leq 0$, a feasible solution achieves values smaller than or equal to zero. Then, the overall constraint violation is measured by summing up all single constraint violations $G(\mathbf{x}) = \sum_{i=1}^{n_c} \max(0, g_i(\mathbf{x}))$, and for constraints of the form $g_i(\mathbf{x}) \geq 0$, the corresponding definition is $G(\mathbf{x}) = \sum_{i=1}^{n_c} \max(0, -g_i(\mathbf{x}))$.

Hence, a positive G indicates a constraint violation. To treat the constraint violation as binary decision problem, we employ the signum function $\text{sgn}(\cdot)$[1] with

$$g(\mathbf{x}) = \text{sgn}(G(\mathbf{x})). \tag{7.1}$$

It holds $g(\mathbf{x}) = 1$, if any constraint g_i is violated and $g(\mathbf{x}) = 0$, if solution $\mathbf{x}$ is feasible. Discretizing the constraint violation is useful for handling the meta-model as binary classification problem. In this chapter, we learn a meta-model $\hat{g}$ of the constraint function with SVMs. We treat the constraint meta-model learning problem as two-class classification problem, i.e., $\hat{g}(\mathbf{x}) = 0$ for a feasible solution $\mathbf{x}$ and $\hat{g}(\mathbf{x}) = 1$ for an infeasible one.

The chapter is structured as follows. First, SVMs are introduced in Sect. 7.2. The meta-model and management mechanism is introduced in Sect. 7.3 for the (1+1)-ES.

---

[1] $\text{sgn}(x) = 1$ for $x > 0$, 0 for $x = 0$, and -1 for $x < 0$.

© Springer International Publishing Switzerland 2016
O. Kramer, *Machine Learning for Evolution Strategies*,
Studies in Big Data 20, DOI 10.1007/978-3-319-33383-0_7

Related work is presented in Sect. 7.4. An experimental study is shown in Sect. 7.5. The chapter closes with conclusions in Sect. 7.6. The benchmark problems are introduced in the appendix.

7.2 Support Vector Machines

Since decades, SVMs belong to the state-of-the art classification algorithms [1]. They have found their way into numerous applications. The variant SVR can be used for regression problems [2]. The idea of SVMs is to learn a separating hyperplane between patterns of different classes. The separating hyperplane should maintain a maximal distance to the patterns of the training set. With the hyperplane, novel patterns $\mathbf{x}'$ can be classified. A hyperplane $\mathbf{H}$ can be described by normal vector $\mathbf{w} = (w_1, \ldots, w_d)^T \in \mathbb{R}^d$ and point $\mathbf{x}_0$ on the hyperplane. For each point $\mathbf{x}$ on $\mathbf{H}$ it holds $\mathbf{w}^T(\mathbf{x} - \mathbf{x}_0) = 0$, as the weight vector is orthogonal to the hyperplane. While defining shift $w_0 = -\mathbf{w}^T \mathbf{x}_0$, the hyperplane definition becomes

$$\mathbf{H} = \{\mathbf{x} \in \mathbb{R}^d : \mathbf{w}^T \mathbf{x} + w_0 = 0\}. \tag{7.2}$$

The objective is to find the optimal hyperplane. This is done by maximizing $1/\|\mathbf{w}\|_2$ corresponding to minimizing $\|\mathbf{w}\|_2$, and the definition of the optimization problem becomes

$$\min \frac{1}{2} \|\mathbf{w}\|_2^2 \tag{7.3}$$

subject to the constraint

$$y_i(\mathbf{w}^T \mathbf{x}_i + w_0) \geq +1, \text{ for all } i = 1, \ldots, N \tag{7.4}$$

Figure 7.1 shows the decision boundary based on a maximizing the margin $\|\mathbf{w}\|_2$. The patterns on the border of the margin are called support vectors. They define the hyperplane that is used as decision boundary for the classification process. Finding $\|\mathbf{w}\|_2$ is an optimization problem, which is the SVM training phase.

The optimization problem is a convex one and can be solved with quadratic programming resulting in the following equation:

$$L_d = \frac{1}{2}(\mathbf{w}^T \mathbf{w}) - \mathbf{w}^T \sum_{i=1}^{N} \alpha_i y_i \mathbf{x}_i - w_0 \sum_{i=1}^{N} \alpha_i y_i + \sum_{i=1}^{N} \alpha_i \tag{7.5}$$

$$= -\frac{1}{2} \mathbf{w}^T \mathbf{w} + \sum_{i=1}^{N} \alpha_i \tag{7.6}$$

$$= -\frac{1}{2} \sum_{i=1}^{N} \sum_{j=1}^{N} \alpha_i \alpha_j y_i y_j \mathbf{x}_i^T \mathbf{x}_j + \sum_{i=1}^{N} \alpha_i \tag{7.7}$$

Fig. 7.1 Decision boundary of an SVM achieved by maximizing margin $1/\|\mathbf{w}\|_2$ and support vector lying on the border of the margin

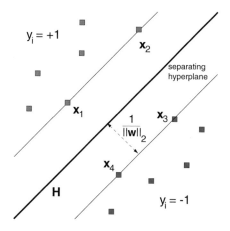

This equation has to be maximized w.r.t. α_i subject to constraints $\sum_{i=1}^{N} \alpha_i y_i = 0$ and $\alpha_i \geq 0$ for $i = 1, \ldots, N$. Maximizing L_d, which stands for the dual optimization problem, can be solved with quadratic optimization methods. The dimensionality of the dual optimization problem depends on the number N of patterns, not on their dimensionality d. The upper bound for the runtime is $\mathcal{O}(N^3)$, while the upper bound for space is $\mathcal{O}(N^2)$.

We get rid of the constraint with a Lagrange formulation [3]. The result of the optimization process is a set of patterns that defines the hyperplane. These patterns are called support vectors and satisfy

$$y_i(\mathbf{w}^T \mathbf{x}_i + w_0) = 1, \qquad (7.8)$$

while lying on the border of the margin, see Fig. 7.1. With any support vector $\mathbf{x}_i$, the SVM is defined as

$$\hat{g}(\mathbf{x}') = \text{sign}(\mathbf{w}^T \mathbf{x}_i + w_0) \qquad (7.9)$$

with $w_0 = y_i - \mathbf{w}^T \mathbf{x}_i$. An SVM that is trained with the support vectors computes the same discriminant function as the SVM trained on the original training set.

For the case that patterns are not separable, slack variables $\xi_i \geq 0$ are introduced that store the deviation from the margin. The optimization problem is relaxed to

$$y_i(\mathbf{w}^T \mathbf{x}_i + w_0) \geq 1 - \xi_i, \qquad (7.10)$$

while the slack variables $\xi_i \leq 0$. The number of misclassifications is $|\{\xi_i > 0\}|$. With the soft error $\sum_{i=1}^{N} \xi_i$, the soft margin optimization problem can be defined as

$$\frac{1}{2}\|\mathbf{w}\|_2^2 + C \cdot \sum_{i=1}^{N} \xi_i \qquad (7.11)$$

Fig. 7.2 Illustration of SVM
learning on XOR data set
consisting of $N = 1000$
patterns. The decision
boundary shows that the
RBF-kernel allows
separating both classes

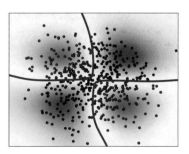

subject to constraints of Eq. 7.10. Penalty factor C serves as regularization parameter
trading off complexity and data misfit, i.e., the number of non-separable patterns.
For the soft margin optimization problem, the Lagrangian formulation is used with
the primal and dual problem transformation.

For handling non-linear data, kernel functions are applied. For kernels, we replace
the inner product computations, which are necessary for solve the SVM optimization problem by the kernel function $K(\mathbf{x}_i, \mathbf{x}_j)$ that compares a similarity between
instances in original input space. Instead of mapping $\mathbf{x}_i$ and $\mathbf{x}_j$ to the abstract feature space and computing the dot product there, the kernel function can directly be
applied in the original space. An explicit mapping to the new space is not necessary,
a concept known as kernel trick. Instead, we use kernel $K : \mathbb{R}^d \times \mathbb{R}^d \to \mathbb{R}$ to replace
dot product in the new space. Besides a linear kernel, an RBF kernel

$$K(\mathbf{x}_i, \mathbf{x}_j) = \exp\left(-\frac{\|\mathbf{x}_i - \mathbf{x}_j\|_2^2}{2\gamma^2}\right) \tag{7.12}$$

with kernel bandwidth γ is often used. Parameter $\gamma > 0$ is usually tuned with grid
search. The matrix of kernel values $\mathbf{K} = [K(\mathbf{x}_i, \mathbf{x}_j)]_{i,j=1}^{N}$ is called kernel or Gram
matrix. In many mathematical models, the kernel matrix is a convenient mathematical
expression. Figure 7.2 shows an example of an SVM with RBF-kernel learning the
XOR data set, see Fig. 7.2.

SVMs are part of the SCIKIT-LEARN package. The following examples illustrate
their use.

- `from sklearn import svm` imports the SCIKIT-LEARN SVM implementation.
- `clf = svm.SVC()` creates a support vector classification (SVC) implementation.
- `clf.fit(X,y)` trains the SVM with patterns X and corresponding labels y.

7.3 Algorithm

In this section, we introduce the constraint meta-model approach, which is based on the (1+1)-ES [4]. We denote the approach as CON-ES. The approach works as follows, see Algorithm 5. In the generational loop, a new offspring solution $\mathbf{x}'$ is generated with recombination and Gaussian mutation. The CON-ES works like the usual (1+1)-ES until the training set size N of patterns and constraint function calls is reached, i.e., a training set $\{(\mathbf{x}_i, g(\mathbf{x}_i))\}_{i=1}^{N}$ is available. In order not to overload the pseudocode, this condition is left out in Algorithm 5. The constraint meta-model $\hat{g}$ is trained with the training set employing cross-validation. As of now, each new solution $\mathbf{x}'$ is first examined on $\hat{g}$ concerning its constraint violation. If feasible, it is evaluated on the real constraint function g. New candidate solutions are generated until meta-model $\hat{g}$ and constraint function g indicate feasibility of $\mathbf{x}'$. The training set $\{(\mathbf{x}_i, g(\mathbf{x}_i))\}_{i=1}^{N}$ is updated in each step and consists of the last N pairs of solutions and their constraint evaluations corresponding to pattern-label pairs. At the end of the loop, step size σ is adapted according to Rechenberg's success rule. A certain balance of labels is required, i.e., $\{(\mathbf{x}_i, g(\mathbf{x}_i))\}_{i=1}^{N}$ is only an appropriate training set, if each class is represented by a similar ratio of labels. The practitioner has to take care for this balance. If large parts of the solution space are constrained, enough feasible solutions have to be generated to make the problem balanced.

Algorithm 5 CON-ES

1: initialize $\mathbf{x}$
2: **repeat**
3:   **repeat**
4:     $\mathbf{z} \sim \sigma \cdot \mathcal{N}(\mathbf{0}, \mathbf{I})$
5:     $\mathbf{x}' = \mathbf{x} + \mathbf{z}$
6:     **if** $\hat{g}(\mathbf{x}') = 0$ **then**
7:       compute $g(\mathbf{x}')$
8:     **end if**
9:   **until** $g(\mathbf{x}') = 0$
10:   last N solutions $\rightarrow \{(\mathbf{x}_i, g(\mathbf{x}_i))\}_{i=1}^{N}$
11:   train $\hat{g}$
12:   evaluate $\mathbf{x}' \rightarrow f(\mathbf{x}')$
13:   **if** $f(\mathbf{x}') \leq f(\mathbf{x})$ **then**
14:     replace $\mathbf{x}$ with $\mathbf{x}'$
15:   **end if**
16:   adapt σ with Rechenberg
17: **until** termination condition

In our CON-ES variant, we use SVMs as constraint meta-models. An SVM is trained on training set $\{(\mathbf{x}_i, g(\mathbf{x}_i))\}_{i=1}^{N}$ with cross-validation and grid search for optimal parameters. A new training might not be necessary in each generation. In the experimental part, we will train the model every 20 generations.

To summarize, based on the constraint model evaluation (in case of infeasibility) or the real constraint function evaluation (in case of feasibility), the feasibility of a new solution $\mathbf{x}'$ is determined. Each time the meta-model $\hat{g}$ predicts infeasibility, a constraint function evaluation on g can be saved.

7.4 Related Work

Since decades of research, many constraint handling methods for evolutionary algorithms have been developed. Methods range from penalty functions that decrease the fitness of infeasible solutions [5] and decoder functions that let the search take place in another unconstrained or less constrained solution space [6] to feasibility preserving approaches that adapt representations or operators [7] to enforce feasibility. Multi-objective approaches treat each constraint as objective that has to be considered separately [8]. For this sake, evolutionary multi-objective optimization methods like non-dominated sorting (NSGA-ii) [9] can be adapted. Penalty functions are powerful methods to handle constraints. A convenient variant is death penalty that rejects infeasible solutions and generates new ones until a sufficient number of feasible candidates are available. A survey on constraint handing for ES gives [10].

Theoretical result on constraint handling and also constraint handling techniques for the CMA-ES [11] are surprisingly rare. For the (1+1)-CMA-ES variant, Arnold and Hansen [12] propose to approximate the directions of the local normal vectors of the constraint boundaries and to use these approximations to reduce variances of the Gaussian distribution in these directions.

We show premature convergence for the Tangent problem [13]. It is caused by dramatically decreasing success probabilities when approximating the constraint boundary. Arnold and Brauer [14] start the theoretical investigation of the (1+1)-ES on the Sphere function with one constraint with a Markov chain analysis deriving progress rates and success probabilities. Similarly, Chotard et al. [15] perform a Markov chain analysis of a $(1, \lambda)$-ES and demonstrate divergence for constant mutation rates and geometric divergence for step sizes controlled with path length control.

Unlike meta-modeling of fitness functions, results on meta-modeling of the constraint boundary are also comparatively rare in literature. Poloczek and Kramer [16] propose an active learning scheme that is based on a multistage model, but employ a linear model with binary search. A pre-selection scheme allows the reduction of constraint function calls. In [17] we combine an adaptive penalty function, which is related to Rechenberg's success rule, with a nearest neighbor fitness and constraint meta-model. Often, the optimal solution of a constrained problem lies in the neighborhood of the feasible solution space. To let the search take place in this region, the adaptive penalty function balances the penalty factors as follows. If less than 1/5th of the population is feasible, the penalty factor γ is increased to move the population into the feasible region

$$\gamma = \gamma/\tau \tag{7.13}$$

with $0 < \tau < 1$. Otherwise, i.e., if more than $1/5$th of the population of candidate solutions is feasible, the penalty is weakened

$$\gamma = \gamma \cdot \tau \tag{7.14}$$

to allow the search moving into the infeasible part of the solution space. The success rate of $1/5$th allows the fastest progress towards the optimal solution.

In [18], Kramer et al. propose a linear meta-model that adapts the covariance matrix of the CMA-ES. The model is based on binary search between the feasible and the infeasible solution space to detect points on the constraint boundary and to define a linear separating hyper-plane. Gieseke and Kramer [19] propose a constraint meta-model scheme for the CMA-ES that employs active learning to select reasonable training points that improve the classifier.

7.5 Experimental Analysis

In this section, we experimentally analyze the CON-ES on two constrained benchmark problems based on the Sphere function. The first is the Sphere function $f(\mathbf{x}) = \mathbf{x}^T\mathbf{x}$ with a linear constraint $g_1(\mathbf{x}) = \sum_{i=1}^{N} x_i \geq 0$ through the optimum $\mathbf{0}$. The optimum lies at the constraint boundary making approximately one half of the population infeasible for isotropic Gaussian mutation. The second constraint is the Tangent problem [12, 13], which is parallel to the previous function but shifted, i.e., $g_1(\mathbf{x}) = \sum_{i=1}^{N} x_i - N \geq 0$. The Tangent results in a constraint function that lies tangential to the contour lines of the fitness function and makes the optimization process a tedious task as the success rates decrease while approximating the optimum.

We employ the following algorithmic settings. The (1+1)-ES uses Gaussian mutation. The first solution is initialized with $\mathbf{x} = (N, \ldots, N)^T$ and the initial step size is $\sigma = 1.0$. Rechenberg's rule employs the setting $\tau = 0.5$ and examines the success rate every 10 generations. We train the SVM with the new training set every 20 iterations and use 5-fold cross-validation and grid search for C and γ with $C, \gamma \in \{10^{-10}, 10^{-9}, \ldots, 10^{10}\}$.

Table 7.1 shows the classification report for the constraint boundary meta-model after 300 generations on the Tangent problem with $d = 10$ dimensions on an arbitrary optimization run. The evaluation is based on one third of the training set. Precision and recall are high for the class of feasible patterns (class 0) and slightly lower for the class of infeasible patterns (class 1). In the following, we will show how the meta-model performs in the course of the constrained optimization process.

Table 7.1 Classification report for constraint meta-model on the tangent problem

Domain	Precision	Recall	F1	Support
0	0.98	0.97	0.97	98
1	0.73	0.80	0.76	10
Avg/total	0.96	0.95	0.95	108

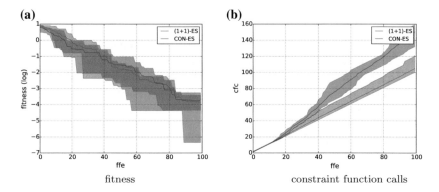

Fig. 7.3 Fitness development and constraint function calls of (1+1)-ES and CON-ES on Sphere with constraint, $N = 2$, and 100 generations

Table 7.2 Experimental comparison of (1+1)-ES without meta-model and CON-ES on the benchmark function set w.r.t. constraint function calls

Problem		(1+1)-ES		CON-ES	
	d	Mean	Dev	Mean	Dev
Sphere	2	74.08	46.46	52.35	30.88
	10	174.65	105.62	156.05	90.30
Tangent	2	87.3	52.76	55.30	33.47
	10	194.25	98.29	175.39	87.30

Figure 7.3 shows the average fitness development of each 30 (1+1)-ES and CON-ES runs and the corresponding number of constraint function savings on the Sphere problem with a constraint through the origin for $N = 2$. Figure 7.3a shows that both algorithms have the same capability to approximate the optimum. The main result is that the CON-ES is able to save a significant number of constraint function calls, see Fig. 7.3b. We observe similar results for the Tangent problem.

In the Table 7.2, we concentrate on the saving capabilities of the SVM constraint meta-model for the two dimensions $N = 2$ and 10 on the benchmark problems. The (1+1)-ES and the CON-ES terminate after 100 generations for $N = 2$ and after 300 generations for $N = 10$. In case of the Sphere with constraint, the optimum is approximated with arbitrary accuracy, while the search stagnates in the vicinity of the optimum in case of the Tangent problem, see [13]. The stagnation is caused by decreasing success rates due to contour lines that become parallel to the constraint boundary when the search approximates the optimum. The results show that the CON-ES saves constraint function evaluations in all cases, i.e., for both dimensions on both problems.

7.6 Conclusions

In practical optimization problems, constraints reduce the size of the feasible solution space and can increase the difficulty of the optimization problem. Many methods have been proposed for handling constraints. Penalty functions belong to the most popular ones. They penalize infeasible solutions by decreasing the fitness function value with a penalty term. The strengths of these penalties are usually controlled deterministically or adaptively.

The constraint handling problem is still not conclusively solved. In particular the reduction of constraint function calls is an important aspect. A promising direction is meta-modeling of the constraint boundary, an approach that is well-known for fitness function surrogates. In this scenario, machine learning models can be applied to learn the feasibility of the solution space based on examples from the past. If constraints deliver a binary value (feasible and infeasible), a classifier can be used as meta-model. We use SVMs to learn the constraints with cross-validation and grid search, while updating the model every 20 generations. For other benchmark problems, these settings have to be adapted accordingly. In the experiments with a (1+1)-ES on a benchmark function set, we demonstrate that significant savings of constraint function evaluations can be achieved. The trade-off between the frequency of meta-model updates and their accuracy has to be balanced accordingly for practical optimization processes.

The combination with fitness meta-models can simultaneously decrease the number of fitness and constraint function evaluations. The interactions between both meta-models have to be taken into account carefully to prevent negative effects.

References

1. Vapnik, V.: The Nature of Statistical Learning Theory. Springer, New York (1995)
2. Smola, A., Vapnik, V.: Support vector regression machines. Adv. Neural Inf. Process. Syst. **9**, 155–161 (1997)
3. Hastie, T., Tibshirani, R., Friedman, J.: The Elements of Statistical Learning. Springer, New York (2009)
4. Beyer, H., Schwefel, H.: Evolution strategies: a comprehensive introduction. Natural Comput. **1**(1), 3–52 (2002)
5. Joines, J., Houck, C.: On the use of non-stationary penalty functions to solve nonlinear constrained optimization problems with GAs. In: Fogel, D.B. (ed.) Proceedings of the 1st IEEE Conference on Evolutionary Computation, pp. 579–584. IEEE Press, Orlando (1994)
6. Koziel, S., Michalewicz, Z.: Evolutionary algorithms, homomorphous mappings, and constrained parameter optimization. Evol. Comput. **7**(1), 19–44 (1999)

7. Schoenauer, M., Michalewicz, Z.: Evolutionary computation at the edge of feasibility. In: Voigt, H.-M., Ebeling, W., Rechenberg, I., Schwefel, H.-P. (eds.) Proceedings of the 4th Conference on Parallel Problem Solving from Nature, PPSN IV 1996, pp. 245–254. Springer, Berlin (1996)
8. Coello, C.A.: Constraint-handling using an evolutionary multiobjective optimization technique. Civil Eng. Environ. Syst. **17**, 319–346 (2000)
9. Deb, K., Agrawal, S., Pratap, A., Meyarivan, T.: A fast elitist non-dominated sorting genetic algorithm for multi-objective optimisation: NSGA-II. In: Proceedings of the 6th International Conference on Parallel Problem Solving from Nature, PPSN IV 2000, pp. 849–858. Paris, France, 18–20 September 2000
10. Kramer, O.: A review of constraint-handling techniques for evolution strategies. Appl. Comput. Int. Soft Comput. **2010**, 185063:1–185063:11 (2010)
11. Hansen, N., Ostermeier, A.: Adapting arbitrary normal mutation distributions in evolution strategies: the covariance matrix adaptation. In: International Conference on Evolutionary Computation, pp. 312–317 (1996)
12. Arnold, D.V., Hansen, N.: A (1+1)-CMA-ES for constrained optimisation. In: Proceedings of the Genetic and Evolutionary Computation Conference, GECCO 2012, pp. 297–304. Philadelphia, PA, USA, 7–11 July 2012
13. Kramer, O.: Premature convergence in constrained continuous search spaces. In: Proceedings of the 10th International Conference on Parallel Problem Solving from Nature, PPSN X 2008, pp. 62–71. Dortmund, Germany, 13–17 September 2008
14. Arnold, D.V., Brauer, D.: On the behaviour of the (1+1)-ES for a simple constrained problem. In: Proceedings of the 10th International Conference on Parallel Problem Solving from Nature, PPSN X 2008, pp. 1–10. Dortmund, Germany, 13–17 September 2008
15. Chotard, A.A., Auger, A., Hansen, N.: Markov chain analysis of evolution strategies on a linear constraint optimization problem. In: Proceedings of the IEEE Congress on Evolutionary Computation, CEC 2014, pp. 159–166. Beijing, China, 6–11 July 2014
16. Poloczek, J., Kramer, O.: Multi-stage constraint surrogate models for evolution strategies. In: Proceedings of the 37th Annual German Conference on AI, KI 2014: Advances in Artificial Intelligence, pp. 255–266. Stuttgart, Germany, 22–26 September 2014
17. Kramer, O., Schlachter, U., Spreckels, V.: An adaptive penalty function with meta-modeling for constrained problems. In: Proceedings of the IEEE Congress on Evolutionary Computation, CEC 2013, pp. 1350–1354 (2013)
18. Kramer, O., Barthelmes, A., Rudolph, G.: Surrogate constraint functions for CMA evolution strategies. In: Proceedings of the 32nd Annual German Conference on AI, KI 2009: Advances in Artificial Intelligence, pp. 169–176. Paderborn, Germany, 15–18 September 2009
19. Gieseke, F., Kramer, O.: Towards non-linear constraint estimation for expensive optimization. In: Proceedings of the Applications of Evolutionary Computation—16th European Conference, EvoApplications 2013, pp. 459–468. Vienna, Austria, 3–5 April 2013

Part IV
Unsupervised Learning

Chapter 8
Dimensionality Reduction Optimization

8.1 Introduction

Optimization in continuous solution spaces can be a tedious task. Various methods have been proposed to allow the optimization in multimodal fitness landscapes. In this chapter, we introduce a further method based on dimensionality reduction. The search takes place in a redundant solution space $\mathbb{R}^{\hat{d}}$ with a higher dimensionality $\hat{d} > d$ than the original solution space. To allow the fitness function evaluation, the high-dimensional candidate solution representation must be mapped back to the solution space with a dimensionality reduction mapping $F : \mathbb{R}^{\hat{d}} \to \mathbb{R}^d$. We call this approach dimensionality reduction evolution strategy (DR-ES). Employing more dimensions than the original solution space specification requires can be beneficial in some solution spaces. The search in the original solution space may be more difficult due to local optima and unflattering solution space characteristics that are not present in the high-dimensional pendants. The assumption of the DR-ES is that the additional dimensions offer a degree of freedom that can better be exploited with the usual evolutionary operators, i.e., intermediate recombination and self-adaptive Gaussian mutation in the case of the (μ, λ)-ES. The mapping from $\mathbb{R}^{\hat{d}}$ to $\mathbb{R}^d$ is computed after a new population has been generated. Any dimensionality reduction method is potentially a good choice for the dimensionality reduction process. We concentrate on PCA, which assumes linearity between variables. But the approach is not restricted to PCA and can be combined with any other point-wise dimensionality reduction approach.

The approach is related to the concept of bloat, which is the phenomenon that parts of the genome do not encode for functional meaningful parts of species. However, bloat is supposed to play an important role in the evolutionary development. Evolution gets the freedom to develop the unused parts that are not under evolutionary pressure to potentially useful functional parts. By introducing more variables for the optimization process than actually necessary to encode the problem, the redundancy is supposed to support overcoming local optima and difficult solution space conditions.

© Springer International Publishing Switzerland 2016
O. Kramer, *Machine Learning for Evolution Strategies*,
Studies in Big Data 20, DOI 10.1007/978-3-319-33383-0_8

This chapter is structured as follows. In Sect. 8.2, we introduce the dimensionality reduction problem. In Sect. 8.3, we briefly present the idea of PCA. The DR-ES approach is introduced in Sect. 8.4. Related work is presented in Sect. 8.5. Section 8.6 presents an experimental analysis of the DR-ES. Conclusions are presented in Sect. 8.7.

8.2 Dimensionality Reduction

Due to an enormous growth of the number of sensors installed in various computer science-oriented domains, Big Data became a relevant research issue in the last years. The sensor resolution has grown steadily leading to the situation of very large data sets with high-dimensional patterns. Dimensionality reduction has an important part to play in solving these Big Data challenges. Putting it in a more formal definition, the dimensionality reduction problem is to find low-dimensional representations $\mathbf{x}_i \in \mathbb{R}^d$ of high-dimensional patterns $\hat{\mathbf{x}}_i \in \mathbb{R}^{\hat{d}}$ for $i = 1, \ldots, N$. To tackle this challenge, numerous dimensionality reduction techniques have been proposed in the past. Famous ones are self-organizing maps introduced by Kohonen [1]. PCA is an excellent method for linear data. It will be introduced in Sect. 8.3. If the data is non-linear, methods like ISOMAP [2] and locally linear embedding (LLE) [3] may be the proper choice. In Chap. 9, we will focus on a non-linear dimensionality reduction method to visualize high-dimensional optimization processes.

Dimensionality reduction approaches perform a mapping F of patterns from a high-dimensional space to a low-dimensional space while maintaining important information. Distance information and neighborhoods should be preserved. For point-wise embeddings, this means that for each high-dimensional pattern, a vector with less dimensions is computed that servers as its low-dimensional pendant. A frequent approach that is known as feature selection concentrates on the search on a smaller number of features. Based on quality measures like entropy or on pure blackbox search, feature selection methods reduce the number of features simply by leaving them out. A good feature selection method finds an optimal combination of features to solve a particular machine learning problem. Due to the curse of dimensionality problem, see Chap. 4, the concentration on a subset of features is an important pre-processing step. Evolutionary search can be employed to select the optimal feature combination. Examples are introduced for various domains, see [4].

8.3 Principal Component Analysis

PCA by Pearson [5, 6] is a method for linear dimensionality reduction. It computes principal components from the training data set. Given a set of $\hat{d}$-dimensional patterns $\hat{\mathbf{x}}_i \in \mathbb{R}^{\hat{d}}$ with $i = 1, \ldots, N$, the objective of PCA is to find a linear manifold of a

Fig. 8.1 Illustration of PCA. The first principal component captures the largest possible variance. Each further component employs the next highest variance and must be orthogonal to the preceding components

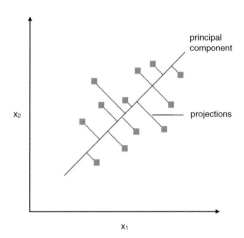

lower dimension $d < \hat{d}$ that captures the most variance of the patterns. Figure 8.1 illustrates the PCA concept. For this sake, PCA computes the covariance matrix of the patterns

$$\mathbf{C} = \frac{1}{N-1} \sum_{i=1}^{N} (\bar{\mathbf{x}} - \hat{\mathbf{x}})(\bar{\mathbf{x}} - \hat{\mathbf{x}})^T \tag{8.1}$$

with mean

$$\bar{\mathbf{x}} = \frac{1}{N} \sum_{i=1}^{N} \hat{\mathbf{x}}_i. \tag{8.2}$$

If $\lambda_1 \geq \cdots \geq \lambda_{\hat{d}}$ are the eigenvalues of the covariance matrix $\mathbf{C}$, and if $\mathbf{e}_1, \ldots, \mathbf{e}_{\hat{d}}$ are the corresponding eigenvectors, we can define a $\hat{d} \times d$-matrix

$$\mathbf{V} = [\mathbf{e}_1, \ldots, \mathbf{e}_d]. \tag{8.3}$$

With this matrix, the mapping $\mathbf{F} : \mathbb{R}^{\hat{d}} \to \mathbb{R}^d$,

$$\mathbf{F}(\hat{\mathbf{x}}_i) = \mathbf{V}^T (\hat{\mathbf{x}}_i - \bar{\mathbf{x}}) \tag{8.4}$$

from data space to the d-dimensional space can be performed. The inverse mapping

$$\mathbf{f}(\mathbf{x}_i) = \mathbf{V}\mathbf{x}_i + \bar{\mathbf{x}} \tag{8.5}$$

back to the $\hat{d}$-dimensional data space is the projection of pattern $\mathbf{x}_i$ onto the linear manifold.

Often, data does not show linear characteristics. Examples are wind time series or image recognition data, where the patterns often live in high dimensions. In this case, non-linear dimensionality reduction methods like ISOMAP and LLE are good choices. The quality of the dimensionality reduction result can be evaluated with measurements that concentrate on the maintenance of neighborhoods like the co-ranking matrix [7], the nearest neighbor classification error for labeled data [8], or by inspection of visualized embeddings.

SCIKIT- LEARN allows an easy integration of PCA with the steps introduced in the following.

- `from sklearn import decomposition` imports the SCIKIT- LEARN decomposition package that contains PCA variants.
- `decomposition.PCA(...).fit_transform(X)` fits PCA to the list of patterns X and maps them to a q-dimensional space. Again, further methods can be employed.

8.4 Algorithm

In this section, we introduce the DR-ES that is based on dimensionality reduction methods that allow the point-wise mapping from $\mathbb{R}^{\hat{d}}$ to $\mathbb{R}^d$ like briefly introduced in the previous section. The DR-ES is a modified self-adaptive population-based (μ, λ)-ES, i.e., it typically employs dominant or intermediate recombination and any form of Gaussian mutation with self-adaptive step size control.

Algorithm 6 shows the DR-ES pseudocode. At the beginning, the candidate solutions $\hat{\mathbf{x}}_1, \ldots, \hat{\mathbf{x}}_\mu \in \mathbb{R}^{\hat{d}}$ and the step sizes $\sigma_1, \ldots, \sigma_\mu$ are initialized. Each solution carries its own step size vector. In the generational loop, λ candidate solutions are produced like in the conventional (μ, λ)-ES. The dimensionality reduction method F, in the experimental part we employ PCA, is applied to the offspring population $\{\hat{\mathbf{x}}_i'\}_{i=1}^\lambda$ resulting in d-dimensional solution candidates $\{\mathbf{x}_i'\}_{i=1}^\lambda$ that are evaluated on f. In the last step, the $\hat{d}$-dimensional μ best solutions $\{\hat{\mathbf{x}}_i\}_{i=1}^\mu$ are selected w.r.t. the fitness the d-dimensional counterparts achieve. The optimization process stops, when a termination condition is met. The ES uses self-adaptive step size control [9], which works as follows. Each solution is equipped with its own step size. It is recombined using the typical recombination operators and is mutated with the log-normal mutation operator

$$\sigma' = \sigma \cdot e^{\tau \mathcal{N}(0,1)} \qquad (8.6)$$

and mutation strength τ. As the step sizes are inherited with the solutions, good step sizes spread in the course of evolution.

To summarize, after application of the dimensionality reduction method, a complete individual is a tuple $(\hat{\mathbf{x}}_i, \sigma_i, \mathbf{x}_i, f(\mathbf{x}_i))$ of a high-dimensional *abstract* solution $\hat{\mathbf{x}}_i$, step size σ_i, which may also be a step size vector, depending on the employed Gaussian mutation type, the candidate solution $\mathbf{x}$ in the original solution space, and its fitness $f(\mathbf{x})$.

Algorithm 6 DR-ES

1: initialize $\hat{\mathbf{x}}_1, \ldots, \hat{\mathbf{x}}_\mu$ and $\sigma_1, \ldots, \sigma_\mu$
2: **repeat**
3:    **for** $j = 1$ **to** λ **do**
4:       recombination($\{\hat{\mathbf{x}}_i\}_{i=1}^\mu$) $\rightarrow \hat{\mathbf{x}}_j'$
5:       recombination($\sigma_1, \ldots, \sigma_\mu$) $\rightarrow \sigma_j'$
6:       log-normal mutation $\rightarrow \sigma_j'$
7:       Gaussian mutation $\rightarrow \hat{\mathbf{x}}_j'$
8:    **end for**
9:    dim. red. (PCA) $F(\hat{\mathbf{x}}_i') \rightarrow \{\mathbf{x}_i'\}_{i=1}^\lambda$
10:   evaluate $\{\mathbf{x}_i\}_{i=1}^\lambda \rightarrow \{f(\mathbf{x}_i)\}_{i=1}^\lambda$
11:   select $\{\hat{\mathbf{x}}_i\}_{i=1}^\mu$ w.r.t. f
12: **until** termination condition

The question how to choose the dimension $\hat{d}$ depends on the problem. In the experimental section, we will experiment with $\hat{d} = 3/2d$, e.g. $\hat{d} = 15$ for $d = 10$. Further method-specific parameters like the neighborhood size of ISOMAP and LLE may have to be chosen according to the optimization problem.

8.5 Related Work

To the best of our knowledge, the evolutionary search in a high-dimensional space with mapping back to the solution space via PCA or other dimensionality reduction methods has not been applied yet.

Related to the idea of employing more dimensions than required for the original problem is related to the discussion mapping from genotypes to phenotypes and of bloat. Often, a function is required that maps the genetic representation of the solution, i.e. the genotype, to the actual solution, i.e. the phenotype. In continuous solution spaces, this is often not required as the genotype can directly be represented as phenotype. A recent contribution to this discussion comes from Simões et al. [10], who employ feed-forward neural networks for defining genotype-phenotype maps of arbitrary continuous optimization problems and concentrate on the analysis of locality and redundancy, and the self-adaptation of the mappings.

The concept of bloat in evolutionary computation is related to our approach of adding additional genes. Bloat are the parts of the genome that are not directly required to encode a solution. This redundancy is supposed to allow walks in solution spaces that overcome local optima and flat fitness landscapes.

Liepins and Vose [11] show that all continuous functions are theoretically easy for EAs given an appropriate representation. But as the mappings are in general unknown and the search space of functions is very large because one has to consider all permutations of mappings, it is difficult to find the optimal representation.

However, representation adaptation on a meta-level is supposed to be attractive in this context.

Wang et al. [12] propose an algorithm that separates the solution space into subspaces by selecting objective variables that are relevant for separate objectives in multi-objective optimization problems. This form of dimensionality reduction is similar to feature selection in machine learning.

To some extend related to the dimensionality reduction idea of the DR-ES is the kernel trick of methods like SVMs. The kernel trick allows implicit operations in a high-dimensional reproducing Hilbert space [13], where, e.g., non-linear data can linearly be separated. The mapping into this kernel feature space is not explicitly computed. Often, only pattern similarities are required that can be computed via the dot product of the original patterns using a kernel function. The DR-ES searches in a high-dimensional abstract space that is mapped back to the low-dimensional solution space.

Decoder functions for constrained solution spaces also share similarities with the DR-ES. Decoder functions map the constrained solution space to an unconstrained, or only box-constrained space, in which the search is easier to carry out. Solutions in this so-called feature space are feasible and the original solutions can easily be repaired. Koziel and Michalewicz [14] employ a homomorphous mapping between the constrained solution space and the one-dimensional cube. Similar, Bremer et al. [15] use kernel functions with support vector description to search for feasible scheduling solutions in the energy domain. To some extend related is the approach by Boschetti [16], who employs LLE to support the evolutionary search with candidate solutions.

Evolutionary computation in dimensionality reduction is also a promising related line of research, e.g., for tuning dimensionality reduction parameters or embedding results with ES. Vahdat et al. [17] proposes to use evolutionary multi-objective algorithms to balance intra-cluster distance and connectedness of clusters. We use a (1+1)-ES to tune clusters of embeddings with ISOMAP, LLE, and PCA by scaling and rotating [18]. In [8], we propose an iterative approach that constructs a dimensionality reduction solution with Gaussian mutation-like steps.

8.6 Experimental Analysis

In the following, the DR-ES is experimentally analyzed. Table 8.1 shows the corresponding results. It shows the mean fitness achieved by a (15,100)-ES and a (15,100)-DR-ES with PCA on Sphere and Rastrigin after 5000 fitness function evaluations. We test the settings $d = 10$ and $d = 20$, while the search takes place in the higher dimensional space $\hat{d} = 15$, and $\hat{d} = 30$, respectively. For example in the case of $d = 20$, the ES searches in the solution space $\mathbb{R}^{30}$, while the PCA maps the best solutions $\hat{\mathbf{x}}_1, \ldots, \hat{\mathbf{x}}_\mu$ back to the original solution space $\mathbb{R}^{20}$. The table shows the medians and corresponding standard deviations of 25 runs for different experimental settings. The results show that the (15,100)-DR-ES with PCA outperforms the (15,100)-ES

Table 8.1 Experimental comparison of (15,100)-ES and (15,100)-DR-ES with PCA on Sphere and Rastrigin

Problem	$\hat{d}/d$	(15,100)-ES		(15,100)-DR-ES		Wilx.
		Median	Dev	Median	Dev	p-value
Sphere	15/10	3.292e-12	4.006e-12	**1.055e-13**	8.507e-14	**1.821e-05**
	30/20	1.931e-06	2.196e-06	**4.960e-08**	4.888e-08	**1.821e-05**
Rastrigin	15/10	2.984	5.65	**1.463e-06**	5.151	**0.0003**
	30/20	56.583	34.43	**0.312**	5.71	**1.821e-05**

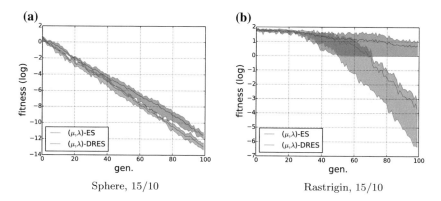

(a) Sphere, $15/10$ · (b) Rastrigin, $15/10$

Fig. 8.2 Comparison of evolutionary runs between (15,100)-ES and (15,100)-DR-ES on **a** the Sphere function and **b** Rastrigin employing $\hat{d} = 15$ and $d = 10$

on Sphere and Rastrigin for all dimensions d and corresponding choices $\hat{d}$. The results are confirmed with the Wilcoxon signed rank-sum test. All values lie below a p-value of 0.05 and are consequently statistically significant. On the Sphere function, the DR-ES achieves slight improvements w.r.t. the median result. This observation is remarkable as the standard (μ, λ)-ES with Gaussian mutation and self-adaptive step size control is known to be a strong optimization approach on the Sphere function. Further, the (μ, λ)-ES fails on Rastrigin, where the DR-ES is significantly superior.

Figure 8.2 compares the evolutionary runs of a (15,100)-ES and a (15,100)-DR-ES on (a) the Sphere function and (b) Rastrigin with $\hat{d} = 15$ and $d = 10$. The plots show the mean, best and worst runs. All other runs lie in light blue and light red regions. The plots show that the DR-ES with PCA is superior to the standard (15,100)-ES. On the Sphere function, the DR-ES is significantly faster, on Rastrigin, it allows convergence in contrast to the stagnating standard ES.

Like observed in Table 8.1, this also holds for the Sphere function. As both ES employ comma selection, the fitness can slightly deteriorate within few optimization steps leading to a non-smooth development with little spikes.

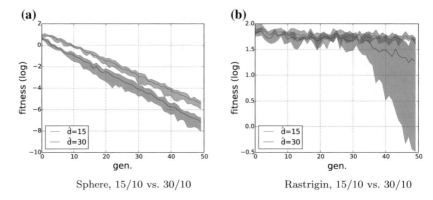

Fig. 8.3 $\hat{d} = 15$ versus $\hat{d} = 30$ comparison on **a** Sphere and **b** Rastrigin with $d = 10$

The question for the influence of $\hat{d}$ arises. In the following, we compare the optimization process between large $\hat{d}$ and moderate $\hat{d}$. Figure 8.3 shows the results of 25 experimental runs on both problems with $d = 10$. The experiments show that a too large solution space deteriorates the capabilities of the DR-ES to approximate the optimum. On the Sphere, the DR-ES with $\hat{d} = 30$ achieves a log-linear development, but is significantly worse than the DR-ES in all phases of the search. On Rastrigin, the DR-ES with lower $\hat{d}$ is able to achieve better approximation capabilities than the approach with $\hat{3} = 50$. The latter does not perform fast runs, but they differ from each other resulting in the large blue area.

8.7 Conclusions

To optimize in multimodal solution spaces, the DR-ES employs a high-dimensional abstract solution space, in which the optimization processes are easier to perform. Dimensionality reduction is used to map the abstract solutions back to the original solution space dimensionality, where they are evaluated on the original fitness function. Dimensionality reduction is the mapping from a high-dimensional space to a low-dimensional space while maintaining important information like pattern neighborhoods and pattern distances. This information is used to compute low-dimensional pendants of the high-dimensional data. The search in the high-dimensional abstract space is performed as usually with continuous recombination and Gaussian mutation. Assisting evolutionary continuous search with dimensionality reduction this kind of way is a novel idea to the best of our knowledge.

For the mapping from the abstract space to the solution space, strong classic dimensionality reduction methods have be used like PCA. Our experimental results show that the DR-ES with PCA is superior to the standard ES without dimensionality reduction. PCA detects the main principal components corresponding to variances in the data. It is based on the computation of the eigenvectors with the largest eigenvalues of the covariance matrix. Obviously, the additional features have an important impact on the search and adding additional degrees of freedoms makes the search easier. In the future, a theoretical analysis will be useful to show the impact of additional features.

References

1. Kohonen, T.: Self-Organizing Maps. Springer (1995)
2. Tenenbaum, J.B., Silva, V.D., Langford, J.C.: A global geometric framework for nonlinear dimensionality reduction. Science **290**, 2319–2323 (2000)
3. Roweis, S.T., Saul, L.K.: Nonlinear dimensionality reduction by locally linear embedding. Science **290**, 2323–2326 (2000)
4. Treiber, N.A., Kramer, O.: Evolutionary feature weighting for wind power prediction with nearest neighbor regression. In: Proceedings of the IEEE Congress on Evolutionary Computation, CEC 2015, pp. 332–337. Sendai, Japan, 25–28 May 2015
5. Jolliffe, I.: Principal component analysis. Springer Series in Statistics. Springer, New York (1986)
6. Pearson, K.: On lines and planes of closest fit to systems of points in space. Philos. Mag. **2**(6), 559–572 (1901)
7. Lueks, W., Mokbel, B., Biehl, M., Hammer, B.: How to evaluate dimensionality reduction? Improving the co-ranking matrix. CoRR (2011)
8. Treiber, N.A., Späth, S., Heinermann, J., von Bremen, L., Kramer, O.: Comparison of numerical models and statistical learning for wind speed prediction. In: Proceedings of the European Symposium on Artificial Neural Networks, ESANN 2015, pp. 71–76 (2015)
9. Beyer, H., Schwefel, H.: Evolution strategies: a comprehensive introduction. Nat. Comput. **1**(1), 3–52 (2002)
10. Simões, L.F., Izzo, D., Haasdijk, E., Eiben, Á.E.: Self-adaptive genotype-phenotype maps: neural networks as a meta-representation. In: Proceedings of the Parallel Problem Solving from Nature, PPSN 2014, pp. 110–119 (2014)
11. Liepins, G.E., Vose, M.D.: Representational issues in genetic optimization. J. Exp. Theor. Artif. Intell. **2**(2), 101–115 (1990)
12. Wang, H., Jiao, L., Shang, R., He, S., Liu, F.: A memetic optimization strategy based on dimension reduction in decision space. Evol. Comput. **23**(1), 69–100 (2015)
13. Aronszajn, N.: Theory of reproducing kernels. Trans. Am. Math. Soc. **68**(3), 404 (1950)
14. Koziel, S., Michalewicz, Z.: Evolutionary algorithms, homomorphous mappings, and constrained parameter optimization. Evol. Comput. **7**(1), 19–44 (1999)
15. Bremer, J., Sonnenschein, M.: Model-based integration of constrained search spaces into distributed planning of active power provision. Comput. Sci. Inf. Syst. **10**(4), 1823–1854 (2013)
16. Boschetti, F.: A local linear embedding module for evolutionary computation optimization. J. Heuristics **14**(1), 95–116 (2008)
17. Vahdat, A., Heywood, M.I., Zincir-Heywood, A.N.: Bottom-up evolutionary subspace clustering. In: Proceedings of the IEEE Congress on Evolutionary Computation, pp. 1–8 (2010)
18. Kramer, O.: Hybrid manifold clustering with evolutionary tuning. In: Proceedings of the 18th European Conference on Applications of Evolutionary Computation, EvoApplications 2015, pp. 481–490. Copenhagen, Denmark (2015)

Chapter 9
Solution Space Visualization

9.1 Introduction

Visualization is the discipline of analyzing and designing algorithms for visual representations of information to reinforce human cognition. It covers many scientific fields like computational geometry or data analysis and finds numerous applications. Examples reach from biomedical visualization and cyber-security to geographic visualization, and multivariate time series visualization. For understanding of optimization processes in high-dimensional solution spaces, visualization offers useful tools for the practitioner. The techniques allow insights into the working mechanisms of evolutionary operators, heuristic components, and their interplay with fitness landscapes.

In particular, the visualization of high-dimensional solution spaces is a task not easy to solve. The focus of this chapter is the mapping with a dimensionality reduction function F from a high-dimensional solution space $\mathbb{R}^d$ to a low-dimensional space $\mathbb{R}^q$ with $q = 2$ or 3 that can be visualized. Modern dimensionality reduction methods like ISOMAP [1] and LLE [2] that have proven well in practical data mining processes allow the visualization of high-dimensional optimization processes by maintaining distances and neighborhoods between patterns. These properties are particularly useful for visualizing high-dimensional optimization processes, e.g., with two-dimensional neighborhood maintaining embeddings.

Objective of this chapter is to show how ISOMAP can be employed to visualize evolutionary optimization runs. It is structured as follows. Section 9.2 gives a short introduction to ISOMAP. In Sect. 9.3, the dimensionality reduction-based visualization approach is introduced. Section 9.4 presents related work on visualizing evolutionary runs. Exemplary runs of ES are shown in Sect. 9.5. Conclusions are drawn in Sect. 9.6.

© Springer International Publishing Switzerland 2016
O. Kramer, *Machine Learning for Evolution Strategies*,
Studies in Big Data 20, DOI 10.1007/978-3-319-33383-0_9

9.2 Isometric Mapping

ISOMAP is based on multi-dimensional scaling (MDS) [3] that estimates the coordinates of a set of points, of which only the pairwise distances δ_{ij} with $i, j = 1, \ldots, N$ and $i \neq j$ are known. ISOMAP uses the geodesic distance, as the data often lives on the surface of a curved manifold. The geodesic distance assumes that the local linear Euclidean distance is reasonable for close neighboring points, see Fig. 9.1. First, ISOMAP determines all points in a given radius ϵ, and looks for the k-nearest neighbors. The next task is to construct a neighborhood graph, i.e., to set a connection to a point that belongs to the k-nearest neighbors and set the corresponding edge length to the geodesic distance. As next step, ISOMAP computes the shortest paths between any two nodes using Dijkstra's algorithm. In the last step, the low-dimensional embeddings are computed with MDS using the previously computed geodesic distances. ISOMAP is based on MDS, which estimates the coordinates of a set of points, while only the distances are known. Let $\mathbf{D} = (\delta_{ij})$ be the distance matrix of a set of patterns with δ_{ij} being the distance between two patterns $\mathbf{x}_i$ and $\mathbf{x}_j$. Given all pairwise distances δ_{ij} with $i, j = 1, \ldots, N$ and $i \neq j$, MDS computes the corresponding low-dimensional representations. For this sake, a matrix $\mathbf{B} = (b_{ij})$ is computed with

$$b_{ij} = -\frac{1}{2}[\delta_{ij}^2 - \frac{1}{N}\sum_{k=1}^{N}\delta_{kj}^2 - \frac{1}{N}\sum_{k=1}^{N}\delta_{ik}^2 + \frac{1}{N^2}\sum_{k=1}^{N}\sum_{l=1}^{N}\delta_{kl}^2]. \qquad (9.1)$$

The points are computed via an eigendecomposition of $\mathbf{B}$ with Cholesky decomposition or singular value decomposition resulting in eigenvalues λ_i and corresponding eigenvectors $\gamma_i = (\gamma_{ij})$. It holds $\sum_{j=1}^{N}\gamma_{ij}^2 = \lambda_i$. The embeddings in a q-dimensional space are the eigenvectors of the q-largest eigenvalues $\hat{\mathbf{x}}_i = \gamma_i\sqrt{\lambda_i}$. Figure 9.2 shows an example of MDS estimating the positions of a set of 10-dimensional candidate positions based on distances between the points. The original points have been generated by a $(1+1)$-ES optimizing the Sphere function. The blue estimated points are

Fig. 9.1 Illustration of geodesic distance measure

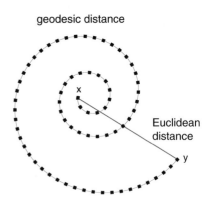

geodesic distance

Euclidean distance

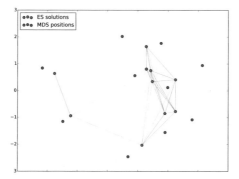

Fig. 9.2 MDS on 10-dimensional patterns generated during ES-based optimization. The *red dots* show the original ES positions of the candidate solutions (first two dimensions), the *blue dots* show the MDS estimates

located close to the original points generated by the ES concerning the first two dimensions. However, the estimation differs from the original positions, as it also considers the remaining eight dimensions.

An example for the application of MDS is depicted in Fig. 9.3. The left part shows the Swiss Roll data in three dimensions, while the right part shows the embedded points computed with MDS. Points that are neighboring in data space are neighboring in the two-dimensional space. This is an important property for our visualization approach.

ISOMAP does not compute an explicit mapping from the high-dimensional to the low-dimensional space. Hence, the embedding of further points is not possible easily. An extension with this regard is incremental ISOMAP [4]. It efficiently updates the solution of the shortest path problem, if new points are added to the data set. Further, the variant solves an incremental eigendecomposition problem with increasing distance matrix.

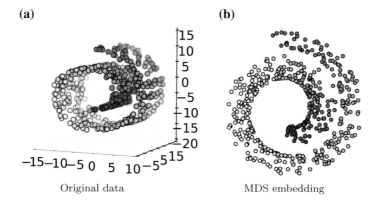

(a)   **(b)**

Original data   MDS embedding

Fig. 9.3 Illustration of dimensionality reduction with MDS: **a** Swiss roll data set and **b** its MDS embedding

Similar to PCA in the previous chapter, SCIKIT-LEARN allows the easy integration of ISOMAP, which is sketched in the following.

- `from sklearn import manifold` imports the SCIKIT-LEARN manifold package that contains ISOMAP, LLE, and related methods.
- `manifold.Isomap(...).fit_transform(X), n_neighbors` fits ISOMAP to the training set of patterns X and maps them to a q-dimensional space with neighborhood size `n_neighbors`, which corresponds to k in our previous description.

In the following, ISOMAP is used to map the solution space to a two-dimensional space that can be visualized. Mapping into a three-dimensional space for visualization in 3d-plots is also a valid approach.

9.3 Algorithm

Dimensionality reduction methods map patterns from a high-dimensional data space, e.g., $\mathbb{R}^d$, to a low-dimensional latent space $\mathbb{R}^q$ with $q \ll d$. Objective of the dimensionality reduction process is to maintain most information like distances and neighborhoods of patterns. This is also discussed in Chap. 8. Again, we treat the solution space $\mathbb{R}^d$ as data space to visualize the optimization process in two dimensions. The dimensionality reduction-based visualization process is shortly sketched in the following algorithm.

Algorithm 7 VIS-ES

1: (1+1)-ES on $f \rightarrow \{\mathbf{x}_i\}_{i=1}^{N}$
2: dim. red. (ISOMAP) $F(\mathbf{x}_i) \rightarrow \{\hat{\mathbf{x}}_i\}_{i=1}^{N}$
3: convex hull of $\{\hat{\mathbf{x}}_i\}_{i=1}^{N} \rightarrow H$
4: generate meshgrid in $H \rightarrow \Gamma$
5: train $\hat{f}$ with $(\hat{\mathbf{x}}_1, f(\mathbf{x}_1)) \dots, (\hat{\mathbf{x}}_N, f(\mathbf{x}_N))$
6: interpolate contour plot $\hat{f}(\gamma) : \forall \gamma \in \Gamma$
7: track search with lines $L_i : \mathbf{x}_i$ to $\mathbf{x}_{i+1}$

Algorithm 7 shows the pseudocode of our visualization approach, which we call VIS-ES in the following. First, the (1+1)-ES performs the optimization run on fitness function f. We visualize the last N generations, which constitute the training set of patterns $\{\mathbf{x}_i\}_{i=1}^{N}$. The patterns are embedded into a two-dimensional latent space ($q = 2$) with ISOMAP and neighborhood size k. Other dimensionality reduction methods can be used as well, e.g., we compare to PCA and LLE in [5].

The embedding process results in a set of low-dimensional representations $\{\hat{\mathbf{x}}_i\}_{i=1}^{N}$ of all candidate solutions. In our visualization approach, embeddings are part of the plots and drawn with circles that are colored according to their fitness.

The next steps of the VIS-ES aim at computing an interpolation of the contour plot based on the fitness in the low-dimensional space. The task is to interpolate from the non-uniformly spaced points to a mesh-grid of points to allow a continuous contour plot of fitness values within the convex hull of embedded candidate solutions. For this sake, first a convex hull H for set $\{\hat{\mathbf{x}}_i\}_{i=1}^{N}$ is computed. A meshgrid, i.e., equally distributed points within the maximum and minimum of coordinates of all $\hat{\mathbf{x}}_i$, serves as basis of the contour lines. All points of this meshgrid within H are basis of the contour line point set Γ. The low-dimensional representations and the corresponding fitness values of their solution space pendants are basis of the training set $(\hat{\mathbf{x}}_1, f(\mathbf{x}_1)), \ldots, (\hat{\mathbf{x}}_N, f(\mathbf{x}_N))$ for the interpolation model $\hat{f}$. This is a regression function that interpolates the surface of fitness landscape f based on the non-uniformly spaced patterns, i.e., $\hat{f}(\gamma)$ is computed for all $\gamma \in \Gamma$. Our implementation of the contour line computation is based on the MATPLOTLIB [6] method GRIDDATA, which can use natural neighbor interpolation based on Delaunay triangulation and linear interpolation. This is also a regression task, but the stack of algorithms is tailored to interpolation in two dimensions.

Last, a set of lines L_i is computed that tracks the evolutionary run by connecting solutions $(\mathbf{x}_i, \mathbf{x}_{i+1})$ of consecutive generations i and $i+1$ with $i = 1, \ldots, N-1$. When employing populations, we recommend to connect the best solutions of consecutive generations like in [5].

9.4 Related Work

An early overview of visualization techniques for evolutionary methods is presented by Pohlheim [7]. He proposes to employ MDS to visualize evolutionary processes. This is demonstrated for visualizing examples of candidate solutions of a population and of non-dominated solutions in multi-objective optimization. MDS-based visualization is appropriate to visualize evolution paths, but lacks from a visualization of the surrounding solution space.

A further early work in this line of research is the visualization of evolutionary runs with self-organizing maps (SOMs) [8]. Also Lotif [9] employs SOMs to visualize evolutionary optimization processes with a concentration on different parameter configurations. SOMs do not compute pattern-by-pattern embeddings like ISOMAP and related methods, but deliver a vector-quantization like placement of weight vectors in data space. These can be used to visualize important parts of data space on the trained map.

Volke et al. [10] introduce an approach for visualizing evolutionary optimization runs that can be applied to a wide range of problems and operators including combinatorial problems. It is based on steepest descent and shortest distance computations in the solution space.

In [11], an adaptive fitness landscape method is proposed that employs MDS. The work concentrates on distance measures that are appropriate from a genetic operator perspective and also for MDS. Masuda et al. [12] propose a method to visualize multi-objective optimization problems with many objectives and high-dimensional decision spaces. The concept is based on distance minimization of reference points on a plane.

Jornod et al. [13] introduce a visualization tool for PSO for understanding PSO processes for the practitioner and for teaching purposes. The visualization capabilities of the solution space are restricted to selecting two of the optimization problem's dimensions at a time, but allow following trajectories and showing fitness landscapes.

Besides visualization, sonification, i.e., the representation of features with sound, is a further way to allow humans the perception of high-dimensional optimization processes, see e.g., Grond et al. [14]. Dimensionality reduction methods can also be directly applied in evolutionary search. For example, Zhang et al. [15] employ LLE in evolutionary multi-objective optimization exploiting the fact that a Pareto set of a continuous multi-objective problem lives in piecewise continuous manifolds of lower dimensionality.

9.5   Experimental Analysis

The VIS-ES is experimentally analyzed in the following. We show the visualization results on our set of benchmark problems. The visualization methods we employ are based on MATPLOTLIB methods, e.g., PLOT and SCATTER. The following figures show the search visualization of a (1+1)-ES with a constant mutation strength of $\sigma = 1.0$ and Rechenberg's step size control on the Sphere function, on the Cigar, on Rosenbrock, and on Griewank.

The visualization is based on the dimensionality reduction of the last $N = 20$ candidate solutions with ISOMAP and neighborhood size $k = 10$. Figure 9.4 shows that the spherical conditions of the optimization process on the 10-dimensional Sphere function is conveniently visualized in the two-dimensional space. On the Cigar function, the walk in the valley can clearly be followed, in case of the Rechenberg variant, see Fig. 9.5, the solution space is significantly narrower. Also on Rosenbrock and on Griewank, the walk in the solution space can appropriately be visualized (Figs. 9.6 and 9.7).

Visualization techniques are difficult to evaluate. To support the observation that the proposed approach maintains high-dimensional properties, we analyze the embeddings w.r.t. the co-ranking matrix measure $E_{NX}(K) \in [0, 1]$ by Lee and Verleysen [16]. It measures the fraction of neighbors of each pattern that occur in a K-neighborhood in data and in latent space. High values for E_{NX} show that the high-dimensional neighborhood relations are preserved in latent space. Our analysis assumes that the maintenance of high-dimensional neighborhoods of evolutionary runs in the low-dimensional plots is an adequate measure for their quality. We analyze the co-ranking matrix measure of the last 100 solutions of 1000 generations of

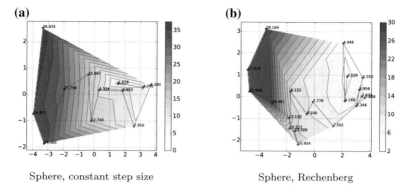

Fig. 9.4 Visualizing of (1+1)-ES run on the Sphere function, $N = 10$ **a** with constant step size, **b** with Rechenberg step size adaptation

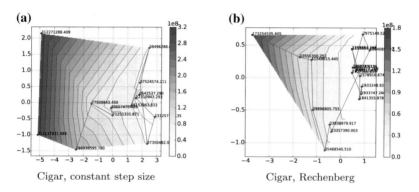

Fig. 9.5 Visualizing of (1+1)-ES run on the Cigar function, $N = 10$ **a** with constant step size, **b** with Rechenberg step size adaptation

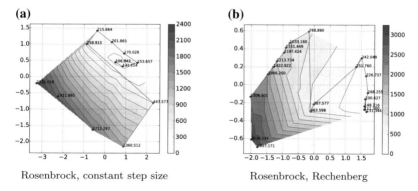

Fig. 9.6 Visualizing of (1+1)-ES run on Rosenbrock with $N = 10$ and **a** with constant step size, **b** with Rechenberg step size adaptation

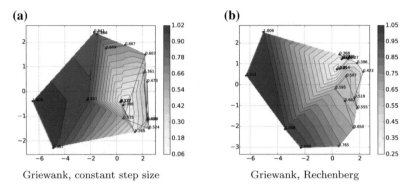

Griewank, constant step size   Griewank, Rechenberg

Fig. 9.7 Visualizing of (1+1)-ES run on the Griewank, $N = 10$ **a** with constant step size, **b** with Rechenberg step size adaptation

Table 9.1 Co-ranking matrix measure of the VIS-ES based on a (1+1)-ES and the COV-ES while embedding 100 solutions of a 20-dimensional evolutionary runs on five benchmark functions

Problem	Sphere	Cigar	Rosenbrock	Rastrigin	Griewank
(1+1)-ES	0.739	0.766	0.811	0.763	0.806
COV-ES	0.806	0.792	0.760	0.829	0.726

the VIS-ES on the Sphere, Cigar, Rosenbrock, Rastrigin, and Griewank with $d = 20$, see Table 9.1. The first line of the table shows the results of the VIS-ES based on ISOMAP with neighborhood size $k = 10$ and the (1+1)-ES, see Algorithm 7. The optimization part of the second line is based on the COV-ES, see Chap. 3. For E_{NX}, we also use a neighborhood size of $k = 10$. The results show that ISOMAP achieves comparatively high values when embedding the evolutionary runs reflecting a high neighborhood maintenance. This result is consistent with our previous analysis of the co-ranking measure for the embedding of evolutionary runs in [5].

9.6 Conclusions

Visualization has an important part to play for human understanding and decision making. The complex interplay between evolutionary runs and multimodal optimization problems let sophisticated visualization techniques for high-dimensional solution spaces become more and more important. In this chapter, we demonstrate how ISOMAP allows the visualization of high-dimensional evolutionary optimization runs. ISOMAP turns out to be an excellent method for maintaining important properties like neighborhoods, i.e., candidate solutions neighboring in the high-dimensional solution space are neighboring in latent space. It is based on MDS and graph-based distance computations. The success of the dimensionality reduction process of the search is demonstrated with the co-ranking matrix measure that indicates the ratio of coinciding neighborhoods in high- and low-dimensional space.

Further dimensionality reduction methods can easily be integrated into this framework. For example, PCA and LLE showed promising results in [5]. The interpolation step for the colorized fitness visualization in the low-dimensional space can be replaced, e.g., by regression approaches like kNN or SVR. Incremental dimensionality reduction methods allow an update of the plot after each generation of the (1+1)-ES. In practice, the visualization can be used to support the evolutionary search in an interactive manner. After a certain number of generations, the search can be visualized, which offers the practitioner the necessary means to evaluate the process and to interact with the search via parameter adaptations.

References

1. Tenenbaum, J.B., Silva, V.D., Langford, J.C.: A global geometric framework for nonlinear dimensionality reduction. Science **290**, 2319–2323 (2000)
2. Roweis, S.T., Saul, L.K.: Nonlinear dimensionality reduction by locally linear embedding. Science **290**, 2323–2326 (2000)
3. Kruskal, J.: Nonmetric multidimensional scaling: a numerical method. Psychometrika **29**, (1964)
4. Law, M.H.C., Jain, A.K.: Incremental nonlinear dimensionality reduction by manifold learning. IEEE Trans. Pattern Anal. Mach. Intell. **28**(3), 377–391 (2006)
5. Kramer, O., Lückehe, D.: Visualization of evolutionary runs with isometric mapping. In: Proceedings of the IEEE Congress on Evolutionary Computation, CEC 2015, pp. 1359–1363. Sendai, Japan, 25–28 May 2015
6. Hunter, J.D.: Matplotlib: a 2d graphics environment. Comput. Sci. Eng. **9**(3), 90–95 (2007)
7. Pohlheim, H.: Multidimensional scaling for evolutionary algorithms—visualization of the path through search space and solution space using sammon mapping. Artif. Life **12**(2), 203–209 (2006)
8. Romero, G., Guervos, J.J.M., Valdivieso, P.A.C., Castellano, F.J.G., Arenas, M.G.: Genetic algorithm visualization using self-organizing maps. In: Proceedings of the Parallel Problem Solving from Nature, PPSN 2002, pp. 442–451 (2002)
9. Lotif, M.: Visualizing the population of meta-heuristics during the optimization process using self-organizing maps. In: Proceedings of the IEEE Congress on Evolutionary Computation, CEC 2014, pp. 313–319 (2014)
10. Volke, S., Zeckzer, D., Scheuermann, G., Middendorf, M.: A visual method for analysis and comparison of search landscapes. In: Proceedings of the Genetic and Evolutionary Computation Conference, GECCO 2015, pp. 497–504. Madrid, Spain, 11–15 July 2015
11. Collier, R., Wineberg, M.: Approaches to multidimensional scaling for adaptive landscape visualization. In: Pelikan, M., Branke, J. (eds.) Proceedings of the Genetic and Evolutionary Computation Conference, GECCO 2010, pp. 649–656. ACM (2010)
12. Masuda, H., Nojima, Y., Ishibuchi, H.: Visual examination of the behavior of emo algorithms for many-objective optimization with many decision variables. In: Proceedings of the IEEE Congress on Evolutionary Computation, CEC 2014, pp. 2633–2640 (2014)
13. Jornod, G., Mario, E.D., Navarro, I., Martinoli, A.: Swarmviz: An open-source visualization tool for particle swarm optimization. In: Proceedings of the IEEE Congress on Evolutionary Computation, CEC 2015, pp. 179–186. Sendai, Japan, 25–28 May 2015
14. Grond, F., Hermann, T., Kramer, O.: Interactive sonification monitoring in evolutionary optimization. In: 17th Annual Conference on Audio Display, Budapest (2011)

15. Zhang, Y., Dai, G., Peng, L., Wang, M.: Hmoeda_lle: A hybrid multi-objective estimation of distribution algorithm combining locally linear embedding. In: Proceedings of the IEEE Congress on Evolutionary Computation, CEC 2014, pp. 707–714 (2014)
16. Lee, J.A., Verleysen, M.: Quality assessment of dimensionality reduction: rank-based criteria. Neurocomputing **72**(7–9), 1431–1443 (2009)

Chapter 10
Clustering-Based Niching

10.1 Introduction

Some optimization problems posses many potential locations of local and global optima. The potential locations are often denoted as basins in solution space. In many optimization scenarios, it is reasonable to evolve multiple equivalent solutions, as one solution may not be realizable in practice. Alternative optima allow the practitioner the fast switching between solutions. Various techniques allow the maintenance of diversity that is necessary to approximate optima in various basins of solution spaces. Such techniques are, e.g., large populations, restart strategies, and niching. The latter is based on the detection of basins and simultaneous optimization within each basin. Hence, niching approaches implement two important steps: (1) the detection of potential niches, i.e., parts of solution space that may accommodate local optima and (2) the maintenance of potentially promising niches to allow convergence of the optimization processes within each niche.

In this chapter, we propose a method to detect multiple locations in solution space that potentially accommodate good local or even global optima for ES [1]. This detection of basins is achieved by sampling in solution space, selecting the best solutions w.r.t. their fitness, and then detecting potential niching locations with clustering. For clustering, we apply DBSCAN [2], which does not require the initial specification of the number of clusters, and k-means that successively repeats cluster assignments and cluster mean computations.

This chapter is structured as follows. Section 10.2 gives a short introduction to clustering concentrating on DBSCAN, k-means, and the DUNN index to evaluate clustering results. Section 10.3 introduces the clustering-based niching concept. Related work is introduced in Sect. 10.4. Experimental results are presented in Sect. 10.5. Last, Sect. 10.6 summarizes the most important findings.

10.2   Clustering

Clustering is the task of grouping patterns without label information. Given patterns $\mathbf{x}_i$ with $i = 1, \ldots, N$, the task is to group them into clusters. Clustering aims at maximizing the homogeneity among patterns in the same cluster and the heterogeneity of patterns in different clusters. Various evaluation criteria for clustering have been presented in the past, e.g., the DUNN index [3] that we use in the experimental part. Some techniques require the specification of the number of clusters at the beginning, e.g., k-means, which is a prominent clustering method. DBSCAN and k-means will shortly be sketched in the following.

DBSCAN [2] is a density-based clustering method. With a user-defined radius `eps` and number `min_samples` of patterns within this radius, DBSCAN determines core points of clusters, see Fig. 10.1. Core points lie within regions of high pattern density. DBSCAN assumes that neighboring core points belong to one cluster. Using the core points, the cluster is expanded and further points within radius `eps` are analyzed. All core points that are reachable from a core point belong to the same cluster. Corner points are points that are reachable from a core point, but that are not core points themselves. Patterns that are neither core points nor corner points are classified as noise.

For comparison, we experiment with the famous clustering method k-means. In k-means, a number k of potential clusters has to be detected before the clustering process begins. First, k cluster centers are randomly initialized. Then, the two steps of assigning patterns to the nearest cluster centers and computing the new cluster centers with the assigned patterns, are iteratively repeated until the movements of the cluster centers fall below a threshold value.

Cluster evaluation measures are often based on inter- and intra-cluster variance. A famous clustering measure is the DUNN index. It computes the ratio between the distance of the two closest clusters and the maximum diameter of all clusters, for an illustration see Fig. 10.2. Let $c(\mathbf{x}_i)$ be a function that delivers the cluster pattern $\mathbf{x}_i$ is assigned to. The minimum distance between all clusters is defined as

$$\delta = \min_{i,j,i \neq j, c(\mathbf{x}_i) \neq c(\mathbf{x}_j)} \|\mathbf{x}_i - \mathbf{x}_j\|_2. \qquad (10.1)$$

Fig. 10.1 Illustration of core and corner points of DBSCAN

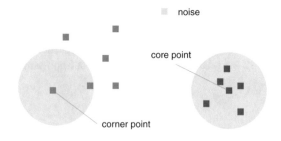

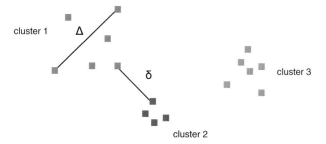

Fig. 10.2 Illustration of DUNN index

The maximal cluster diameter is defined as

$$\Delta = \max_{i,j,i\neq j, c(\mathbf{x}_i)=c(\mathbf{x}_j)} \|\mathbf{x}_i - \mathbf{x}_j\|_2. \tag{10.2}$$

The DUNN index is defined as δ/Δ and has to be maximized, i.e., small maximal cluster diameters and large minimal cluster distances are preferred. The DUNN index is useful for our purpose, as small niches and large distances between niches are advantageous for the maintenance of niches during the optimization process.

The application of DBSCAN and k-means in SCIKIT- LEARN has already been introduced in Chap. 5 and is only shortly revisited here.

- DBSCAN(eps=0.3, min_samples=10).fit(X) is an example, how DBSCAN is accessed for a pattern set X, also illustrating the use of both density parameters.
- KMeans(n_clusters=5).fit(X) is the corresponding example for k-means assuming 5 clusters.

10.3 Algorithm

In this section, we introduce the clustering-based niching approach. Algorithm 8 shows the pseudocode, which is denoted as NI-ES in the following. In the initialization step, the objective is to detect all niches. For this sake, the approach first samples λ' candidate solutions randomly with uniform distribution in the whole feasible solution space. This initial phase targets at exploring the solution space for the detection of basins. For this sake, each solution $\mathbf{x}_1, \ldots, \mathbf{x}_{\lambda'}$ has to be evaluated w.r.t. fitness function f. Among the λ' candidates the $N = \varphi \cdot \lambda'$ best solutions are selected. The proper choice of rate $\varphi \in (0, 1)$ depends on the structure of the optimization problem, e.g., on the number and size of basins as well as on the dimensionality d of the problem.

Algorithm 8 NI-ES

1: initialize $\mathbf{x}_1, \ldots, \mathbf{x}_{\lambda'}$ (random, uniform dist.)
2: evaluate $\{\mathbf{x}_i\}_{i=1}^{\lambda'} \rightarrow \{f(\mathbf{x}_i)\}_{i=1}^{\lambda'}$
3: select N best solutions
4: cluster $\mathbf{x}_1, \ldots, \mathbf{x}_N \rightarrow C$
5: **for** cluster in C **do**
6:    cluster center $\rightarrow$ initial solution $\mathbf{x}$
7:    intra-cluster variance $\rightarrow$ initial step size σ
8:    (1+1)-ES until termination condition
9: **end for**

In the next step, the remaining candidate solutions are clustered. From the selection process, basins turn out to be agglomerations of patterns in solution space that can be detected with clustering approaches. The result of the clustering process is an assignment of the N solutions to k clusters C.

For the optimization within each niche, i.e., in each cluster of C, an initial step size for the Gaussian mutation has to be determined from the size of the basins. The step size should be large enough to allow fast search within a niche, but small enough to prevent their unintentional leaving. We propose to employ the intra-cluster variance as initial step size σ. With the center of each niche as starting solution $\mathbf{x}$, k (1+1)-ES begin their search in each niche until their termination conditions are reached. This step can naturally be parallelized.

First, the process concentrates on the detection of potential niches. For this sake, a random initialization with uniform distribution is performed in solution space. The number of candidate solutions during this phase must be adapted to the dimension d of the problem. In the experimental part, we will focus on the curse of dimensionality problem. The trade-off in this step concerns the number of patterns. A large number improves the clustering result, i.e., the detection of niches, but costs numerous potentially expensive fitness function evaluations.

10.4 Related Work

Niching is a method for multimodal optimization that has a long tradition [4]. Shir and Bäck [5] propose an adaptive individual niche radius for the CMA-ES [6]. Pereira et al. [7] integrate nearest-better clustering and other heuristic extensions into the CMA-ES. Similar to our approach, their algorithm applies an exploratory initialization phase to detect niches.

Sadowski et al. [8] propose a clustering-based niching approach that takes into account linkage-learning and that is able to handle binary and real-valued objective variables including constraint handling. Preuss et al. [9] take into account properties like size relations, basins sizes, and other indicators for the identification of niches. For clustering, nearest-better clustering and Jarvis-Patrick clustering are used.

For multi-objective optimization, we employ DBSCAN to detect and approximate equivalent Pareto subsets in multi-objective optimization [10]. This approach uses explorative cluster detection at the beginning, but tries to explore niches during the usual optimization runs. Also Bandaru and Deb [11] concentrate on niching for multi-objective optimization arguing that dominance and diversity preservation inherently cause niching.

Niching is a technique that is also applied in other heuristics. Biswas et al. [12] focus on the maintenance of solutions in each niche with an information-sharing approach that allows parallel convergence in all niches. Their experimental analysis concentrates on differential evolution, but the approach can be applied to PSO and other heuristics as well.

In EDAs, restricted tournament replacement is a famous niching concept, and is applied, e.g. in [13]. It randomly samples a set of solutions from the solution space, searches the closest in the population w.r.t. a genotypic distance measure like the Euclidean distance in $\mathbb{R}^d$, and replaces the solution in the tournament set, if its fitness is worse than the fitness of the close solution. Cheng et al. [14] compare PSO algorithms with different neighborhood structures defining the particle groups for the PSO update step on the CEC 2015 single objective multi-niche optimization benchmark set. They report that the ring neighborhood structure performs best on the test set. For variable mesh optimization, Navarro et al. [15] propose a general niching framework. Niching concepts find various applications, e.g., in learning classifier systems [16] or interestingly for clustering data spaces [17].

10.5   Experimental Analysis

In the following, we experimentally analyze the NI-ES by visualizing the clustering results. We analyze the behavior of DBSCAN and k-means for clustering-based niching under different initial sampling conditions. Further, we analyze the maintenance of niches when optimizing with (1+1)-ES in each niche.

Initially, $\lambda' = 1000$ solutions are randomly sampled with uniform probability in the interval $[0, 2]^d$, i.e., $[0, 2]^2$ for the visualized examples, see Fig. 10.3 for DBSCAN and Fig. 10.4 for k-means. Both figures compare the clustering results w.r.t. ratios $\varphi = 0.7$ and $\varphi = 0.3$ of selected solutions. DBSCAN uses the settings `eps = 0.3` and `min_samples = 5`, while k-means gets the known number of clusters, i.e. $k = 4$. The plots show that both algorithms are able to detect all clusters with exception of DBSCAN for ratio $\varphi = 0.7$ of selected solutions. This is due to the closeness of the four clusters, as the ratio of selected solutions is too large. Note, that k-means has the advantage of knowing the correct number of clusters.

In the following, we analyze the properties of the cluster result after clustering the exploratory solution set. Table 10.1 shows the number of correctly identified clusters, the intra- and inter-variance, and the DUNN index for $\lambda' = 1000$ and 10000 patterns, and ratios $\varphi = 0.1$ and 0.05 of selected solutions. The table shows that all parameter choices found the complete set of niches (4 out of 4). The inter-cluster variances

(a)   **(b)**

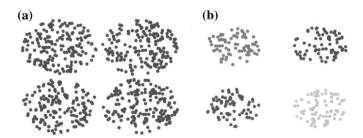

Fig. 10.3 Clustering results of DBSCAN (eps $= 0.3$ and min_samples $= 5$) on the niche benchmark problem for $\varphi = 0.7$ and 0.3 corresponding to $N = 700$ and $N = 300$. Patterns with same colors belong to the same clusters. **a** DBSCAN $\varphi = 0.7$. **b** DBSCAN $\varphi = 0.3$

(a)   **(b)**

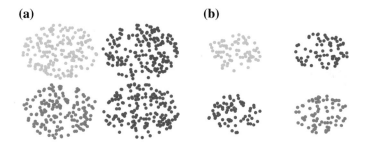

Fig. 10.4 Clustering results of k-means ($k = 4$) on the niche benchmark problem for $\varphi = 0.7$ and $\varphi = 0.3$. **a** k-means $\varphi = 0.7$. **b** k-means $\varphi = 0.3$

Table 10.1 Analysis of number of detected clusters, intra-, inter-cluster variance, and DUNN index for DBSCAN for various data set sizes N and ratios φ on the niche benchmark problem with $d = 2$

λ'	φ	#	Intra	Inter	DUNN
1000	0.1	4/4	0.0096	0.2315	1.6751
1000	0.05	4/4	0.0029	0.2526	3.7522
10000	0.1	4/4	0.0078	0.2517	1.8239
10000	0.05	4/4	0.0037	0.2508	3.0180

are larger than the intra-cluster variances. Further, the results show the intra-cluster variances shrink with higher selection ratio φ, as the patterns are missing that a further away from the niches' optima. This also results in a larger DUNN index value, as the diameters of the clusters are smaller and the clusters are further away from each other.

Now, we combine the explorative niche detection with the evolutionary optimization process employing a (1+1)-ES in each niche. After initialization of **x** with the cluster center that belongs to its niche, the evolutionary loop begins with the intra-cluster variance as initial step size σ. Figure 10.5 shows the optimization process of multiple (1+1)-ES with Rechenberg's step size control and $\tau = 0.5$. In each niche, an independent (1+1)-ES optimizes for 200 generations. The plots show the mean,

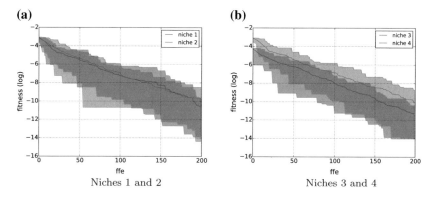

Fig. 10.5 Fitness development of 50 runs (mean, best, and worst runs) of four (1+1)-ES **a** in niches 1 and 2 and **b** in niches 3 and 4 running for 200 generations

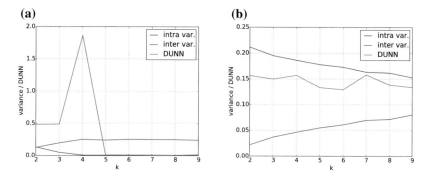

Fig. 10.6 Analysis of intra-cluster variance, inter-cluster variance, and DUNN index w.r.t. the number of clusters k when clustering with k-means **a** for $d = 2$ and **b** for $d = 10$

best, and worst fitness developments of 50 runs on a logarithmic scale. Our approach uses the mean of each cluster as initial solution and the square root of the intra-cluster variance as initial step size. The figures show that the optima are approximated logarithmically linear in each niche. An analysis of the approximated optima shows that the initial choice of step sizes is appropriate as no run of the (1+1)-ES leaves its assigned niche, and the logarithmically linear development starts from the early beginning.

Last, we analyze the dependency of the number of clusters when clustering with k-means on the three parameters intra-cluster variance, inter-cluster variance, and the DUNN index. Figure 10.6 shows the results when sampling with $\lambda = 10000$ points and rate $\varphi = 0.1$, i.e., $N = 1000$ for $d = 2$ on the left hand side and for $d = 10$ on the right hand side. The figures show that the inter-cluster variance increases with the number of clusters, while the intra-cluster variance decreases. In case of $d = 2$, a clear DUNN index maximum can be observed for $k = 4$. Due to the curse of dimensionality problem, the proper choice of cluster numbers does not show a similar DUNN index optimum for $d = 10$ like we observe for $d = 2$.

10.6   Conclusions

In multimodal optimization, the challenge is to detect multiple local optima. Niching is a technique that allows the detection and maintenance of multiple local and global optima at the same time during the evolutionary optimization process. This allows the practitioner to choose among alternative solutions if necessary.

In this chapter, we propose an approach of uniformly sampling patterns in solution space, fitness-based selection, and subsequent clustering for detection of potential niches. The basis for the detection of clusters is an exhaustive scan for basins in the initialization phase. This is performed with uniform random sampling in the allowed solution space. For the experimental test problem, we concentrate on the interval [0, 2] for each dimension. After the selection of the best solutions w.r.t. their fitness function values, clusters of potential basins appear. They are detected with k-means that affords the specification of the number of clusters before its run and DBSCAN that requires a definition of density parameters. The experimental analysis reveals that k-means and DBSCAN have proven to detect niches with proper parameter settings concerning the number of potential niches or solution space densities.

In the optimization phase, the approach to estimate step size σ with the intra-cluster variance is a successful concept to maintain the niches while optimizing with a $(1+1)$-ES. In these scenarios, we do not observe that a cluster is left or forgotten. The optimization in niches can naturally be parallelized and further blackbox optimizers can be applied.

References

1. Beyer, H., Schwefel, H.: Evolution strategies—a comprehensive introduction. Nat. Comput. **1**(1), 3–52 (2002)
2. Ester, M., Kriegel, H.-P., Sander, J., Xu, X.: A density-based algorithm for discovering clusters in large spatial databases with noise. In: Proceedings of the 2nd International Conference on Knowledge Discovery and Data Mining, (KDD 1996), pp. 226–231. AAAI Press (1996)
3. Dunn, J.: A fuzzy relative of the isodata process and its use in detecting compact well-separated clusters. J. Cybern. **3**(3), 32–57 (1973)
4. Eiben, A.E., Smith, J.E.: Introduction to Evolutionary Computing. Springer, Berlin (2003)
5. Shir, O.M., Bäck, T.: Niche radius adaptation in the CMA-ES niching algorithm. In: Proceedings of the 9th International Conference on Parallel Problem Solving from Nature, PPSN IX 2006, pp. 142–151. Reykjavik, Iceland, 9–13 Sept 2006
6. Hansen, N., Ostermeier, A.: Adapting arbitrary normal mutation distributions in evolution strategies: the covariance matrix adaptation. In: International Conference on Evolutionary Computation, pp. 312–317 (1996)
7. Pereira, M.W., Neto, G.S., Roisenberg, M.: A topological niching covariance matrix adaptation for multimodal optimization. In: Proceedings of the IEEE Congress on Evolutionary Computation, CEC 2014, pp. 2562–2569. Beijing, China, 6–11 July 2014
8. Sadowski, K.L., Bosman, P.A.N., Thierens, D.: A clustering-based model-building EA for optimization problems with binary and real-valued variables. In: Proceedings of the Genetic and Evolutionary Computation Conference, GECCO 2015, pp. 911–918. Madrid, Spain, 11–15 July 2015

9. Preuss, M., Stoean, C., Stoean, R.: Niching foundations: basin identification on fixed-property generated landscapes. In: Proceedings of the 13th Annual Genetic and Evolutionary Computation Conference, GECCO 2011, pp. 837–844. Dublin, Ireland, 12–16 July 2011

10. Kramer, O., Danielsiek, H.: Dbscan-based multi-objective niching to approximate equivalent pareto-subsets. In: Proceedings of the Genetic and Evolutionary Computation Conference, GECCO 2010, pp. 503–510. Portland, Oregon, USA, 7–11 July 2010

11. Bandaru, S., Deb, K.: A parameterless-niching-assisted bi-objective approach to multimodal optimization. In: Proceedings of the IEEE Congress on Evolutionary Computation, CEC 2013, pp. 95–102. Cancun, Mexico, 20–23 June 2013

12. Biswas, S., Kundu, S., Das, S.: Inducing niching behavior in differential evolution through local information sharing. IEEE Trans. Evol. Comput. **19**(2), 246–263 (2015)

13. Hsu, P., Yu, T.: A niching scheme for edas to reduce spurious dependencies. In: Proceedings of the Genetic and Evolutionary Computation Conference, GECCO 2013, pp. 375–382. Amsterdam, The Netherlands, 6–10 July 2013

14. Cheng, S., Qin, Q., Wu, Z., Shi, Y., Zhang, Q.: Multimodal optimization using particle swarm optimization algorithms: CEC 2015 competition on single objective multi-niche optimization. In: Proceedings of the IEEE Congress on Evolutionary Computation, CEC 2015, pp. 1075–1082. Sendai, Japan, 25–28 May 2015,

15. Navarro, R., Murata, T., Falcon, R., Hae, K.C.: A generic niching framework for variable mesh optimization. In: Proceedings of the IEEE Congress on Evolutionary Computation, CEC 2015, pp. 1994–2001. Sendai, Japan, 25–28 May 2015

16. Vargas, D.V., Takano, H., Murata, J.: Self organizing classifiers and niched fitness. In: Proceedings of the Genetic and Evolutionary Computation Conference, GECCO 2013, pp. 1109–1116. Amsterdam, The Netherlands, 6–10 July 2013

17. Sheng, W., Chen, S., Fairhurst, M.C., Xiao, G., Mao, J.: Multilocal search and adaptive niching based memetic algorithm with a consensus criterion for data clustering. IEEE Trans. Evol. Comput. **18**(5), 721–741 (2014)

Part V
Ending

Chapter 11
Summary and Outlook

11.1 Summary

ES are famous blackbox optimization algorithms. The variants with Gaussian mutation are tailored to continuous optimization problems. In the last fifty years, they have grown to strong optimization algorithms. Many applications have proven that ES are excellent methods for practical optimization problems. Decades of intensive research and numerous heuristic extensions allow optimization in multimodal, constrained, dynamic, and multi-objective solution spaces. Moreover, ES have their regular tracks on important conferences like GECCO and CEC.

This book introduces various kinds of ways to support ES with machine learning approaches. The (1+1)-ES is a comparatively simple ES and serves as basis of this book to demonstrate the possibilities and potentials of machine learning for evolutionary computation. A training set of patterns or pattern-label pairs serve as basis of the underlying machine learning methods. The following list summarizes these approaches.

- The COV-ES estimates the covariance matrix of a population to improve the success probability of Gaussian mutations. It requires the following step:

  - Estimate the covariance matrix with Ledoit-Wolf with $\{\mathbf{x}_i\}_{i=1}^{N} \rightarrow \mathbf{C}$.

  The Ledoit-Wolf covariance matrix estimator turns out to be successful in the experimental analysis, e.g., it outperforms the variant with empirical covariance matrix estimation.

- The MM-ES learns meta-models of the fitness function for saving potentially expensive evaluations. It is based on nearest neighbor regression and a training set consisting of the last successful solutions similar to the COV-ES. It performs the steps:

  - Construct training set of the last N solutions $\rightarrow \{(\mathbf{x}_i, f(\mathbf{x}_i))\}_{i=1}^{N}$ and
  - train $\hat{f}$, e.g., with kNN and k chosen via cross-validation.

© Springer International Publishing Switzerland 2016
O. Kramer, *Machine Learning for Evolution Strategies*,
Studies in Big Data 20, DOI 10.1007/978-3-319-33383-0_11

- The CON-ES saves constraint function evaluations with a classifier that is trained on solution and constraint function evaluations. The variant analyzed in this book uses an SVM meta-model. The CON-ES training is mainly based on the two steps:

  - Build a training set of the last N solutions $\rightarrow \{(\mathbf{x}_i, g(\mathbf{x}_i))\}_{i=1}^{N}$ and
  - train $\hat{g}$, e.g., with SVM and cross-validation to obtain optimal SVM parameters.

- The DR-ES optimizes in a high-dimensional solution space employing dimensionality reduction to map to the original space. The search in the abstract solution space of higher dimensions appears to be easier than the search in the original one. The key step is:

  - Perform dimensionality reduction $F(\hat{\mathbf{x}}_i') \rightarrow \{\mathbf{x}_i'\}_{i=1}^{\lambda}$, e.g., with PCA.

  Here, $\hat{\mathbf{x}}_i'$ is the set of offspring solutions and F is the dimensionality reduction mapping.

- The VIS-ES allows a mapping from high-dimensional solution spaces to two dimensions for visualizing optimization runs in a fitness landscape. The maintenance of neighborhoods and distances of solutions allow the practitioner the visualization of important processes that take place in the high-dimensional solution space. The machine learning steps of the VIS-ES are:

  - Run the (1+1)-ES on $f \rightarrow \{\mathbf{x}_i\}_{i=1}^{N}$ to obtain a training set and
  - reduce its dimensionality $F(\mathbf{x}_i) \rightarrow \{\hat{\mathbf{x}}_i\}_{i=1}^{N}$, e.g., with ISOMAP.

- The NI-ES detects niches in multimodal solution spaces with an exploratory initialization phase of sampling and clustering. Afterwards, initialization parameters are estimated and the NI-ES optimizes in each niche. Here, the important steps are:

  - Initialize $\mathbf{x}_1, \ldots, \mathbf{x}_{\lambda'}$ randomly with uniform distribution,
  - select N best solutions to obtain the training set, and
  - cluster $\mathbf{x}_1, \ldots, \mathbf{x}_N \rightarrow C$.

The concepts illustrate the potentials of machine learning for ES. They all have in common that a training set of patterns is managed during the optimization process. This training set is often based on a subset of solutions. The pattern distributions often change during the optimization process. For example, when approximating optima in continuous solution space, the distributions become narrower. This domain shift can be handled by restricting the training set to a subset of last solutions. Furthermore, different types of label information are used. The COV-ES and the DR-ES do not use label information. The VIS-ES does not use labels for the dimensionality reduction process, but employs the fitness for interpolation and colorization of the low-dimensional space. The MM-ES uses the fitness values of solutions as labels for training the meta-model, while the CON-ES does the same with the constraint function values for the constraint surrogate.

11.2 Evolutionary Computation for Machine Learning

Related to *machine learning for evolution strategies* is the line of research *evolutionary computation for machine learning*. Many difficult optimization problems arise in machine learning, e.g., when training a machine learning model to fit the observations. Noise in the data, ill-conditioned pattern distributions, and many other potentially difficult conditions can complicate the optimization problem of fitting the model. There are numerous optimization problems in machine learning, for which EAs turn out to be excellent choices.

Complicated data space conditions, complex pattern distributions, and noise induce difficult optimization problems. They leave large space for robust optimization heuristics like evolutionary algorithms. Typically, the following five problem classes arise:

- Tuning of parameters. For example, the parameters of an SVM [1, 2] can be tuned with evolutionary algorithms instead of grid-search. Although the solution space is not too large in many tuning scenarios, evolution is often faster in finding appropriate settings. The combination of cross-validation with evolutionary search yields robust tuned machine learning models.
- Balancing models. Balancing is a special variant of the parameter tuning task. It allows the consideration of two or more objectives, typically prediction error and model complexity. The flexibility of a complex model often has to be paid with a long runtime and computational complexity, respectively. We use evolutionary multi-objective optimization for balancing ensemble classifiers w.r.t runtime and accuracy properties [3] with non-dominated sorting [4]. A comprehensive survey of multi-objective optimization for data mining is presented by Mukhopadhyay et al. [5, 6].
- Pre-processing like feature selection and feature space scaling. The choice of relevant features and their scaling is an important task, as most methods are based on the computation of pattern distances and densities. In this line of research, we scale the feature space of wind power times series data [7] in kNN-based prediction achieving a significant win in accuracy.
- Evolutionary learning. The learning strategy itself that adapts the machine learning model can be an evolutionary method. Evolutionary construction of a dimensionality reduction solution [8, 9] and learning classifier systems are examples for machine learning algorithms with evolutionary optimization as learning basis.
- Post-optimization. The post-optimization of the learning result often allows final improvements. After the main optimization work is done with other algorithms, evolutionary methods often achieve further improvements by fine-tuning. The tuning and rotation of submanifolds with a $(1+1)$-ES in the hybrid manifold clustering approach we introduce in [10] is an example for effective evolutionary post-optimization.

Figure 11.1a shows an example for balancing classifiers of kNN and decision tree ensembles, which is taken from our analysis in [3]. The figure shows the errors and

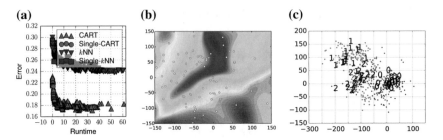

Fig. 11.1 Examples for evolutionary computation for machine learning. **a** Balancing ensemble classifiers with multi-objective evolutionary algorithms, **b** data space reconstruction error of UKR on the Digits data set motivating the employment of evolutionary search, **c** incremental supervised embeddings on the Digits data set. **a** Balancing ensemble classiers on `make_classification`, from [3]. **b** Data space reconstruction error of UKR on Digits, from [12]. **c** Incremental supervised embeddings of Digits, from [13]

the corresponding runtimes of kNN (single kNN), kNN ensembles, the decision tree CART [11], and a CART ensemble on an artificial benchmark classification data set generated with the SCIKIT-LEARN method `make_class`. The evolutionary optimization process is based on NSGA-ii [4]. The runtime is influenced by the feature diversity, i.e., the evolutionary process selects an appropriate set of features. We can observe a Pareto front of solutions that has been evolved by the EA for both ensemble variants. From this Pareto front, the practitioner can choose among alternative solutions, similar to the solutions that have been generated in a niching optimization process. The Pareto front of the decision trees is located lower left of the Pareto front of the nearest neighbor ensembles in objective space, i.e., the decision tree ensembles outperform the nearest neighbor ensembles on `make_class`. On other data sets, the nearest neighbor ensembles turn out to be superior, see the experimental analysis in [3].

An example for evolutionary learning is the optimization unsupervised regression models [14] for dimensionality reduction. Unsupervised regression is based on the idea to optimize the positions of representations in the low-dimensional space w.r.t. to the objective to minimize the regression error when mapping from this space to the high-dimensional space. The high-dimensional patterns are treated as labels, i.e., each dimension is considered separately. The deviation of this mapping and the set of patterns is called data space reconstruction error. Figure 11.1b shows a sample plot of the data space reconstruction error induced by unsupervised kernel regression (UKR) [15] on the Digits data set when embedding on pattern. The plot is taken from [12], where we optimize the data space reconstruction error with evolutionary methods. It shows that the fitness landscape for one pattern is not easy with local optima. In the approach we also alternate gradient descent and evolutionary search to overcome local optima.

Another powerful learning strategy is the incremental construction of a dimensionality reduction solution. In this setting, the low-dimensional points are added to the dimensionality reduction result pattern by pattern. We analyze numerous variants to search for the optimal position of each pattern in each step. In [16], we employ evolutionary binary search to construct a discrete solution, in [8], we use PSO to determine continuous low-dimensional positions. The position in the low-dimensional space of the closest embedded pattern is used as starting point, while the best found latent position serves as attractor during the search. A fast and effective way is to generate solutions based on sampling multiple times from the Gaussian distribution. From an evolutionary perspective, this method is a $(1,\lambda)$-ES running for only one generation.

A further variant for labeled data minimizes the MSE when mapping from low-dimensional space to label space [13]. The result is an alternative set of patterns that minimizes the training error on the new pattern set. Figure 11.1c shows the low-dimensional representation of the incremental supervised dimensionality reduction method on the Digits data set. It is based on Gaussian distributed candidates with the $(1,\lambda)$-ES like operator taking into account label information. Not surprisingly, the use of label information leads to improved separations of patterns with different labels.

The list of successful applications of evolutionary approaches for machine learning is steadily growing. This shows the potentials of both lines of research and lets us expect many interesting research contributions in the future.

11.3 Outlook

The algorithmic variants introduced in this book improve the $(1+1)$-ES and support it, e.g., by visualizing the optimization processes. The methods and concepts can serve as blueprints for variants that are instantiated with other optimization methods and machine learning approaches.

For example, we employ a population-based (μ, λ)-ES in Chap. 8. The (μ, λ)-ES can also be used instead of the $(1+1)$-ES for all other variants. In such settings, the question arises how to choose the training set size. It can have the size μ of the parental population like in case of the DR-ES. Also for covariance matrix estimation of the COV-ES, the choice $N = \mu$ is reasonable. For fitness and constraint meta-models, the training set size should usually be larger to improve the model quality. Of course, this depends on the problem type and dimension. Further, the convergence processes can change the distributions and usually afford a training set adaptation during the run.

The introduced concepts can also be applied to PSO. The covariance matrix estimation mechanism is not applicable to the standard PSO equation, which does not use the Gaussian distribution, but uniform random numbers for scaling the particle directions. However, the fitness function and constraint function surrogates can easily be integrated into PSO-based search. For example, when evaluating the particles' fitness, the prediction of the meta-model can first be checked before the real fitness function is used for promising candidate solutions. The dimensionality reduction process can also be adapted, the particles fly in the abstract space of higher dimensionality, and a training set has to be collected that can be mapped to the original search space dimensionality. The same holds for the visualization mapping that maps the solution space to a two-dimensional printable space.

Interesting future developments comprise the application and adaption of latest developments in machine learning like deep learning [17, 18]. It will surely be interesting to observe the developments in this line of research in the near future.

References

1. Glasmachers, T., Igel, C.: Maximum likelihood model selection for 1-norm soft margin svms with multiple parameters. IEEE Trans. Pattern Anal. Mach. Intell. **32**(8), 1522–1528 (2010)
2. Stoean, C. Stoean, R.: Support Vector Machines and Evolutionary Algorithms for Classification—Single or Together?, volume 69 of Intelligent Systems Reference Library Springer (2014)
3. Oehmcke, S., Heinermann, J., Kramer, O.: Analysis of diversity methods for evolutionary multi-objective ensemble classifiers. In: Proceedings of the 18th European Conference on Applications of Evolutionary Computation, EvoApplications 2015, pp. 567–578. Copenhagen, Denmark, 8–10 April 2015
4. Deb, K., Agrawal, S., Pratap, A., Meyarivan, T.: A fast elitist non-dominated sorting genetic algorithm for multi-objective optimisation: NSGA-II. In: Proceedings of the 6th International Conference on Parallel Problem Solving from Nature, PPSN VI 2000, pp. 849–858. Paris, France, 18–20 Sept 2000
5. Mukhopadhyay, A., Maulik, U., Bandyopadhyay, S., Coello, C.A.C.: A survey of multiobjective evolutionary algorithms for data mining: part I. IEEE Trans. Evol. Comput. **18**(1), 4–19 (2014)
6. Mukhopadhyay, A., Maulik, U., Bandyopadhyay, S., Coello, C.A.C.: Survey of multiobjective evolutionary algorithms for data mining: part II. IEEE Trans. Evol. Comput. **18**(1), 20–35 (2014)
7. Treiber, N.A., Kramer, O.: Evolutionary feature weighting for wind power prediction with nearest neighbor regression. In: Proceedings of the IEEE Congress on Evolutionary Computation, CEC 2015, pp. 332–337. Sendai, Japan, 25–28 May 2015
8. Kramer, O.: A particle swarm embedding algorithm for nonlinear dimensionality reduction. In: Proceedings of the 8th International Conference on Swarm Intelligence, ANTS 2012, pp. 1–12. Brussels, Belgium, 12–14 Sept 2012
9. Kramer, O.: Dimensionality Reduction with Unsupervised Nearest Neighbors, volume 51 of Intelligent Systems Reference Library. Springer (2013)
10. Kramer, O.: Hybrid manifold clustering with evolutionary tuning. In: Proceedings of the 18th European Conference on Applications of Evolutionary Computation, EvoApplications 2015, pp. 481–490. Copenhagen, Denmark (2015)
11. Breiman, L., Friedman, J.H., Olshen, R.A., Stone, C.J.: Classification and Regression Trees. Wadsworth (1984)
12. Lückehe, D., Kramer, O.: Leaving local optima in unsupervised kernel regression. In: Proceedings of the 24th International Conference on Artificial Neural Networks and Machine Learning, ICANN 2014, pp. 137–144. Hamburg, Germany, 15–19 Sept 2014
13. Kramer, O.: Supervised manifold learning with incremental stochastic embeddings. In: Proceedings of the 23rd European Symposium on Artificial Neural Networks, ESANN 2015, pp. 243–248. Bruges, Belgium (2015)

14. Meinicke, P., Klanke, S., Memisevic, R., Ritter, H.: Principal surfaces from unsupervised kernel regression. IEEE Trans. Pattern Anal. Mach. Intell. **27**(9), 1379–1391 (2005)
15. Klanke, S., Ritter, H.: Variants of unsupervised kernel regression: general cost functions. Neurocomputing **70**(7–9), 1289–1303 (2007)
16. Kramer, O.: On evolutionary approaches to unsupervised nearest neighbor regression. In: Proceedings of the Applications of Evolutionary Computation—EvoApplications 2012: Evo-COMNET, EvoCOMPLEX, EvoFIN, EvoGAMES, EvoHOT, EvoIASP, EvoNUM, EvoPAR, EvoRISK, EvoSTIM, and EvoSTOC, pp. 346–355. Málaga, Spain, 11–13 April 2012
17. Bengio, Y.: Learning deep architectures for AI. Found. Trends Mach. Learn. **2**(1), 1–127 (2009)
18. Deng, L., Yu, D.: Deep learning: Methods and applications. Found. Trends Signal Process. **7**(3–4), 197–387 (2014)

Appendix A
Benchmark Functions

The results presented in this book are based on experimental computer experiments on artificial benchmark functions. In the following, we give an overview of employed test functions that are also famous problems in literature. The optimization problems are continuous, i.e., the search takes place in $\mathbb{R}^d$. The Sphere function, see Fig. A.1a, is the problem of minimizing

$$f(\mathbf{x}) = \mathbf{x}^T \mathbf{x} \tag{A.1}$$

with a minimum at $\mathbf{x}^* = (0, \dots, 0)^T$ and $f(\mathbf{x}^*) = 0$. The equation corresponds to the term $f(\mathbf{x}) = \sum_{i=1}^{N} x_i^2$. The Sphere is unimodal and symmetric.

We used two constrained variants. The first employs a linear function, i.e., $g_1(\mathbf{x}) = \sum_{i=1}^{N} x_i \geq 0$ that passes through the origin. The second constrained variant is the Tangent problem with constraint:

$$g_1(\mathbf{x}) = \sum_{i=1}^{N} x_i - t > 0, \qquad t \in \mathbb{R} \tag{A.2}$$

For $N = k$ and $t = k$, the new minimum is $\mathbf{x}^* = (1, \dots, 1)^T$ with $f(\mathbf{x}^*) = k$. It employs a linear function that lies tangential to the contour lines of the fitness function, but suffers from decreasing success probabilities when approximating the optimum.

Similar to the Sphere function is the Cigar function, see Fig. A.1b. Its first dimension is squared, while all further dimensions are weighted with a large constant:

$$f(\mathbf{x}) = x_1^2 + 10^6 \cdot \sum_{i=2}^{N} x_i^2 \tag{A.3}$$

Again, the minimum is at $\mathbf{x}^* = (0, \dots, 0)^T$ with $f(\mathbf{x}^*) = 0$. In Fig. A.1b, this constant is chosen as 10 (instead of 10^6) to illustrate the global structure.

© Springer International Publishing Switzerland 2016
O. Kramer, *Machine Learning for Evolution Strategies*,
Studies in Big Data 20, DOI 10.1007/978-3-319-33383-0

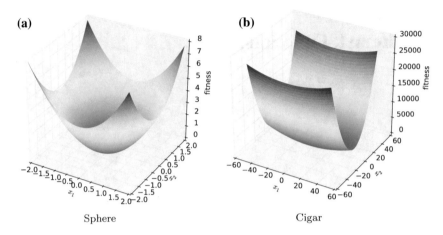

Fig. A.1 Fitness landscape of **a** the Sphere function and **b** the Cigar function

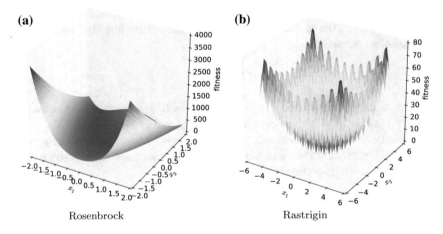

Fig. A.2 Fitness landscape of **a** the Rosenbrock function and **b** the Rastrigin function

The Rosenbrock function, see Fig. A.2a, is defined as minimizing

$$f(\mathbf{x}) = \sum_{i=1}^{N-1} \left((100(x_i^2 - x_{i+1})^2 + (x_i - 1)^2 \right) \tag{A.4}$$

with a minimum at $\mathbf{x}^* = (1, \dots, 1)^T$ with $f(\mathbf{x}^*) = 0$. For higher dimensions, the function has a local optimum. It is non-separable, scalable, and employs a very narrow valley from local optimum to global optimum.

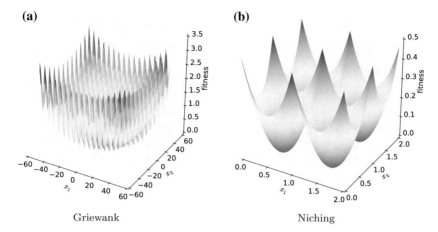

(a) Griewank

(b) Niching

Fig. A.3 Fitness landscape of **a** the Griewank function and **b** the niching function, i.e., the Sphere-based benchmark function with modulo-operator

The Rastrigin function, see Fig. A.2b, is the problem of minimizing

$$f(\mathbf{x}) = \sum_{i=1}^{N} \left(x_i^2 - 10\cos(2\pi x_i) + 10 \right) \tag{A.5}$$

with a minimum at $\mathbf{x}^* = (0, \ldots, 0)^T$ and $f(\mathbf{x}^*) = 0$. It is multimodal, separable, and has a large number of local optima.

Griewank, see Fig. A.3a, is the problem of minimizing

$$f(\mathbf{x}) = \sum_{i=1}^{N} \frac{x_i^2}{4000} - \prod_{i=1}^{N} \cos\left(\frac{x_i}{\sqrt{i}}\right) + 1 \tag{A.6}$$

with a minimum at $\mathbf{x}^* = (0, \ldots, 0)^T$ and $f(\mathbf{x}^*) = 0$. It is also multimodal and non-separable.

The niche benchmark problem, see Fig. A.3b, that is analyzed in the experimental part of Chap. 10 is based on the Sphere function $f(\mathbf{x}) = \mathbf{x}^T\mathbf{x}$. It is modified in the way that for the first d' dimensions x_i with $i = 1, \ldots, d'$ of $\mathbf{x}$ the arguments are taken modulo 1 corresponding to

$$f(\mathbf{x}) = \sum_{i=1}^{d'} (x_i \bmod 1 - 0.5)^2 + \sum_{i=d'+1}^{d} (x_i)^2 \tag{A.7}$$

with the bound constraint $x_i \in [0, 2]$ for all $i = 1, \ldots, d$. With the bound constraint, we get $2^{d'}$ optima. In the experiments of Chap. 10, we choose $d' = 2$ resulting in 4 local/global optima.

Index

© Springer International Publishing Switzerland 2016
O. Kramer, *Machine Learning for Evolution Strategies*,
Studies in Big Data 20, DOI 10.1007/978-3-319-33383-0

Printed in the United States
By Bookmasters